EXPLODING STARS
AND
INVISIBLE PLANETS

EXPLODING STARS AND INVISIBLE PLANETS

THE SCIENCE OF WHAT'S OUT THERE

FRED WATSON

Columbia University Press
New York

Columbia University Press
Publishers Since 1893
New York   Chichester, West Sussex
cup.columbia.edu

First published in Australia by NewSouth,
an imprint of UNSW Press Ltd.
Copyright © 2019 Fred Watson
All rights reserved

Library of Congress Cataloging-in-Publication Data

Names: Watson, Fred, 1944– author.
Title: Exploding stars and invisible planets: the science of what's out there / Fred Watson.
Description: New York: Columbia University Press, [2019] | Includes index.
Identifiers: LCCN 2019025491 (print) | LCCN 2019025492 (e-book) | ISBN 9780231195409
(cloth) | ISBN 9780231540056 (e-book)
Subjects: LCSH: Astronomy--Popular works.
Classification: LCC QB44.3.W378 2019 (print) | LCC QB44.3 (e-book) | DDC 520--dc23
LC record available at https://lccn.loc.gov/2019025491
LC e-book record available at https://lccn.loc.gov/2019025492

∞
Columbia University Press books are printed on permanent and durable acid-free paper.
Printed in the United States of America

Design: Josephine Pajor-Markus
Cover design: Julia Kushnirsky
Cover image: Swirls of dust illuminated by light echoes from variable star V838 Monocerotis.
Courtesy of NASA, ESA, and H. Bond (STScI).

CONTENTS

Prologue: Astronomy at Large   vii

EARTH AND SPACE

CHAPTER 1   Restless Earth: The way of the world   3

CHAPTER 2   The terminator: A user's guide to nightfall   10

CHAPTER 3   Starring citizen science: Research by the people   23

CHAPTER 4   Catch a falling star: Meteors, meteorites and space dust   33

CHAPTER 5   Radio silence: The quietest place in the world   47

CHAPTER 6   The off-planet economy: Doing business in space   56

CHAPTER 7   Moonstruck: Where did our satellite come from?   70

PLANETARY EXPLORATIONS

CHAPTER 8   Telescope troubles: Astronomers in court   83

CHAPTER 9   Space bugs: Rules for planetary protection   99

CHAPTER 10   Climate change: What happened to Mars?   109

CHAPTER 11   Not our Planet B? Colonising Mars   115

CHAPTER 12   Ringing in the changes: The vanishing rings of Saturn   124

CHAPTER 13   Stormy weather: Weird worlds of the Saturnian system   132

CHAPTER 14   Stalking an invisible planet: The search for Planet Nine   142

THE UNIVERSE AT LARGE

CHAPTER 15 Nature's barcode: A user's guide to light   153

CHAPTER 16 Reverberations: Exploding stars and light echoes   167

CHAPTER 17 Signals from the unknown: The fast radio
            burst mystery   180

CHAPTER 18 Eye of the storm: Black holes inside and out   188

CHAPTER 19 Through gravity's lens: The curious matter of
            dark matter   200

CHAPTER 20 Ripples in space: Probing the birth of the Universe   212

CHAPTER 21 Unrequited love: Is anyone there?   226

Acknowledgments   236
Index   238

★

PROLOGUE:
ASTRONOMY AT LARGE

The story begins in darkness, literally: a total eclipse of the Sun. And no ordinary eclipse, either, but the first one ever to be broadcast live on television, documenting the Moon's shadow as it progressed through France, Italy and the former Yugoslavia. The BBC's doyen of astronomy communicators, Patrick Moore, was stationed atop snow-covered Mount Jastrebac in what is now Serbia, and provided a live commentary on the progress of the eclipse. Such as it was – my recollection is that there was a lot of cloud about. There was plenty for him to talk about, though, including the oxen that had hauled the outside broadcast equipment up to the summit. As predicted, they nodded off to sleep in the darkness of totality. Rather to Patrick's chagrin, the producer immediately turned on floodlights to allow viewers to see the dozing animals. Not really what you want in the middle of an eclipse.

Watching all this as a 16-year-old on a black-and-white TV in the cold of a Yorkshire winter's morning, sleep was the furthest thing from my mind. There and then, I resolved to become an astronomer. Perhaps it was the live action of scientists using telescopes to probe the secrets of the Sun's corona – its outer atmosphere, whose mechanisms we still don't fully understand nearly six decades later. Or perhaps it was Patrick's skill in telling viewers exactly what was going on, when half the time he couldn't really see anything because of the cloud.

Eight and a half years later, with sixth form and university behind me, another TV programme held me in thrall, this time showing a chap called Neil Armstrong walking on the Moon. By then I was working for a renowned British company that built large telescopes for astronomers – including several I'd use later in my career. My job at the time was to fabricate the mirrors for a new space telescope that would survey the Universe in ultra-violet radiation. Because the company was truly ancient – well over a century old – it was accustomed to building telescopes so weighty they were measured in tonnes. That didn't really translate into satellite equipment, and we had all kinds of problems producing the lightweight mirrors required. Nevertheless, my telescope eventually flew aboard a robotic spacecraft with the unglamorous name of TD1A.

OVER THE YEARS, I BUILT UP A STORE OF EXPERIENCE IN many different branches of astronomy and space science, which eventually propelled me into the uncharted realm of management. So, for almost two decades, I was the Astronomer-in-Charge of what was then called the Anglo-Australian Observatory, or AAO – a bi-national venture that operated two telescopes at Siding Spring Observatory in north-western New South Wales. One of them, the 3.9-metre Anglo-Australian Telescope, remains the largest optical (visible-light) telescope on Australian soil.

In 2010, however, in a thoroughly polite and terribly British way, the UK pulled out of the deal, leaving the Australian government to run what then became the Australian Astronomical Observatory, or AAO. And eight years later, in a further deal

involving a strategic partnership with a major European observatory, the AAO became part of the university sector. The telescopes at Siding Spring would now be operated by the Australian
National University and the instrument building division in
Sydney rebadged as – wait for it – Australian Astronomical Optics,
or AAO. One thing you can say for the AAO is that it knows how
to save money on logos. The same one has sufficed since 1991, and
still proudly proclaims the organisation's heritage.

So, what happened to the Astronomer-in-Charge amid all
these reorganisations? The structure of the observatory had
changed, and my management role had metamorphosed into
education and outreach with a generous sprinkling of airtime on
the national broadcaster, the ABC (Australian Broadcasting Corporation). So the AAO's parent government department decided
they'd quite like to hang onto me after the 2018 transition. That
suited me very well, of course, since my addiction to communicating astronomy and space science to anyone who would listen
remained undiminished.

But what would my new job be called? Someone suggested
that if my title was tweaked to Astronomer-at-Large, we'd only
have to change four letters on the office door. We sniggered at the
criminal overtones. 'Police have issued a warning that there's an
astronomer at large. Do not attempt approach or capture.' But
Australia's Minister for Industry, Science and Technology, the
Honourable Karen Andrews, really liked the idea, and who was
I to argue?

AS ASTRONOMER-AT-LARGE, I GET TO ENGAGE WITH
researchers all over the world, and relish keeping up to date with
their work so I can bring it to the Australian public on-air. Not to
mention anyone else who's interested. Over the years, it has been

my privilege to select a broad and quirky range of topics for fun radio segments, ranging from asteroid-mining to astrophysics and from Galileo to gravitation. And what a great trove to include in a book.

So *Exploding Stars and Invisible Planets* is based on the 'Astronomer-at-Large's Pick' of seriously interesting astronomy topics. It's an opportunity to bring you some of the less well-known stories from the frontiers of astronomy and space science. Stuff you might not have thought about before, together with a look at what the future might hold. Some of the fields of study featured here are developing very quickly, so what you have is a snapshot of our knowledge as of the middle of 2019.

Let's take a look at what you're going to find within these pages. We'll start on our own planet with some earthy topics that don't normally find their way into books about astronomy and space. The focus of part one is the magical interface between humans, our planet and the sky. Where else would you find an exposition of the glories of sunset, for example, or the place of citizen science in astronomy? Not to mention the way our planet is continuously being bombarded by ancient debris left over from the Solar System's formation. We'll also have a look at the burgeoning space economy, before taking a trip to our marvellous Moon in search of its origins. How appropriate, given that I'm writing this in the fiftieth anniversary year of the first moonwalk.

I mentioned Galileo a minute ago, and we'll revisit his crimes at the start of the section exploring the Solar System. The history of astronomy gives wonderful insights into the science and, as you'll find, its controversies don't stop with Galileo. Then, coming right up to date, we find that planetary studies are conducted today with more than half an eye on the prospects of discovering life elsewhere in the Sun's family. Several chapters in this section follow that trail, before we wind up with the latest on the hunt for a mysterious planet on the outer fringes of the Solar System.

And then we'll turn to the wider Universe. Here, we cover a pretty complete selection of the hot topics in contemporary astrophysics. Light echoing around the cosmos, uncanny radio bursts, the mechanics of black holes – and not one, but two varieties of enigmatic stuff permeating the Universe that make astronomers look silly because we don't know what they are. And just to settle everyone down at the end, we'll take a romantic look at unrequited love. Make sure you have a good supply of tissues handy.

I can't tell you what a privilege it has been to write about all the wonderful research being carried out, as well as relating a little of the curious and occasionally comical history of our science. Honestly, it's nearly as good as watching an eclipse.

EARTH
AND
SPACE

CHAPTER 1

★

RESTLESS EARTH: THE WAY OF THE WORLD

Suppose you could come with me to a place that is typical of the Universe. A location that experiences the average conditions found throughout the whole of space. Where would we be? On the surface of an alien planet, perhaps, luxuriating among exotic plants and strange, colourful creatures? Or close to the brilliant churning atmosphere of a hot star, with tortuous magnetic fields funnelling lethal bursts of plasma towards us? Falling into a black hole? Or just – nowhere?

It's the last of these that is closest to the truth. A typical place in the Universe is empty, cold and dark. And nothing in our experience can quantify just how empty, cold and dark it is. If you're lucky, you might find one atom of hydrogen in the volume of space normally taken up by 15 adults – a cubic metre. The temperature you'd experience is 2.7 degrees above absolute zero, or –270 °C. That's cold. And, to your unaided eyes, the darkness is complete.

But don't worry – I'm not going to leave you here. From this typical spot, we can move at the speed of light towards a place that, after 100 million years or so of travel, will reveal itself to our eyes as a gigantic disc of stars, dust and glowing gas in space. It's set among a handful of other swirls of light now becoming visible, but this one is special enough to have a name, and is known as the Milky Way Galaxy. As we approach it at light speed, another 100 000 years brings us into its suburbs, now visible as a shimmering haze of stars and pink clouds of hydrogen, with dusty patches between them. And setting our sights on one unassuming star brings us to a curious collection of planets – four small rocky ones and four big gassy ones, with a lot of small debris roaming between them. The third planet out from the star looks a bit unusual, with blue and white colouring interrupted by occasional patches of reddish-brown. Mind you, it's nothing compared to the weird one with the rings around it.

As we finally touch down on solid ground, we find there could hardly be a better place in which to check out our home planet. We're in the wilderness of southern Darmaraland in Namibia, surrounded by house-sized granite boulders flushed pink as the rising Sun adds its own hue to the iron-rich stone. There's precious little vegetation in this desert landscape, and the restless history of the Earth's surface is clearly revealed in the tumbling spine of mountains before us. They speak of a time 130 million years ago, when molten rock spilled from gigantic fractures in the supercontinent of Gondwana as it broke apart. Its remnants are present-day Africa, South America, Antarctica, Australia and the Indian subcontinent.

The realisation that plate-like segments of the Earth's crust, or lithosphere – which ranges from 60 to 250 kilometres thick – are in a state of constant movement was one of the great triumphs of mid 20th-century geophysics. It was a theory whose time had come, and half a century of scepticism was ending when I was a pimply teenager at school in the 1960s. New mathematical modelling of heat flow in the Earth's mantle (the underlying layer of soft rock that extends some 2900 kilometres below the surface) had shown that upwelling plumes of viscous rock could, indeed, drive breakneck motion in continental plates. Think lava lamps, and you'll see what I mean. And yes, I know – 'breakneck' is an adjective seldom used in geology, but it's justified in this case: the African and South American plates separate at 2 to 3 centimetres per year, roughly the speed at which your fingernails grow.

It's the ever-widening boundary between these two plates that forms the Mid-Atlantic Ridge, a submarine feature that extends almost from pole to pole, and breaks the ocean surface only in the youthful volcanic landscape of Iceland. While we're used to hearing about tectonic activity in places like Japan, Sumatra and New Zealand – where plates converge, often with disastrous seismic consequences – it is in Iceland where the dynamics of our planet are perhaps at their most visible. As the island is unrelentingly torn in half, volcanic activity is commonplace.

To the best of our knowledge, Earth is unique in the Solar System in having plate tectonics – at least in the present era. And its vigorous geology has spawned a rich chemistry on and near the surface, stimulating a wealth of pre-biotic reactions – and, some three billion years ago, the emergence of living organisms. Today, life blazes forth in all its myriad forms: even here in Darmaraland, where noble desert elephants epitomise its ability to adapt to the most adverse conditions. And we all know the ultimate consequence of biological adaptation. It has produced the most complex entity known in the Universe – the extraordinary brain of *Homo sapiens*.

JUST AFTER SUNSET TONIGHT, THE CLEAR NAMIBIAN SKY will bring a feast of Solar System celebrities. Deep in the western twilight, the planet Venus will herald giant Jupiter high above, while Saturn vies for prominence in the north-east. But it's the slender crescent close to Jupiter that will grab everyone's attention. At this phase, the Moon's disc is bathed in Earthshine (sunlight reflected from the full Earth in the lunar sky), and its dusky surface is faintly visible between the sunlit horns of the crescent. Earthshine has a practical scientific use, explored a few years ago by Canadian and French astronomers. It is the sum total of daylight from the whole Earth – oceans, landmasses, clouds and ice-caps. Cities, towns, golf courses and beer gardens. Everything – and, by analysing it using the rainbow spectroscopy described in chapter 15, astronomers can look for signs of life on our own planet in a trial of the technique's effectiveness for observing the planets of other stars in the future.

Most of us take the Moon for granted, but its gravitational influence has probably been pivotal in the evolution of life on Earth. For example, ocean tides may have been important in animal life gradually migrating from a water environment to dry land, as the twice-daily flooding of the coastline provided a conducive environment. And, more fundamentally, the 'flywheel' effect of a large moon orbiting Earth is believed to have stabilised our planet's axial tilt, keeping it within a whisker or so of its current value of 23.5 degrees. That has promoted stable climatic seasons favourable to biological evolution, and contrasts with a planet such as Mars, which is known to have experienced large changes of tilt (up to 20 degrees) over relatively short timescales (approximately 100 000 years).

Our planet has two other attributes that have assisted in the evolution of life. One is its nickel-iron core, whose diameter of 6970 kilometres is rather more than half that of the planet. At the centre of the molten outer core is a 2440-kilometre-diameter solid metal sphere under extreme pressure, and at a temperature recently estimated to be 6500 °C. Convection currents in the liquid core give rise to the Earth's magnetic field and generate the magnetosphere, a protective barrier that effectively shields the planet's surface and atmosphere from destructive bombardment by the solar wind. This is no benign zephyr, but an energetic stream of electrically charged subatomic particles from the Sun.

At irregular intervals, the dynamo-like interaction between the Earth's solid and liquid cores causes the geomagnetic field to fall in intensity, and occasionally to reverse. It's possible this might occur again within the next couple of thousand years, given the 10 to 15 per cent decline in magnetic field strength that has been observed since measurements began in the mid-19th century. So – no geomagnetism equals no magnetosphere, and a threat to life? Not quite – the interaction of the Sun's magnetic field with the Earth's metallic core induces magnetism that will at least partially protect our fragile environment.

And finally, the Earth's atmosphere provides more than just the air we breathe. A large fraction of the 50 tonnes of meteoritic material that bombards our planet daily (at velocities between 11 and 72 kilometres per second) is harmlessly vaporised 95 kilometres or so above the Earth's surface. Objects that make it into the lower atmosphere or survive long enough to hit the ground as meteorites are relatively rare. You'll read all about them in chapter 4. And, back on the subatomic scale of inbound material, the atmosphere substantially reduces the radiation dose of galactic cosmic rays at the Earth's surface. Dangerous particles again. Clearly, without our planet's blanket of air, we would be at the mercy of a decidedly hostile environment.

But the atmosphere is constantly in a delicate balancing act between monumental geophysical forces. Crucial to its long-term stability is the greenhouse effect of carbon dioxide, and its circulation between the mantle and the air we breathe. This provides a natural thermostat that depends on plate tectonics. When converging tectonic plates collide, the oceanic plate slides under its continental neighbour in an action known as subduction. But with it goes a

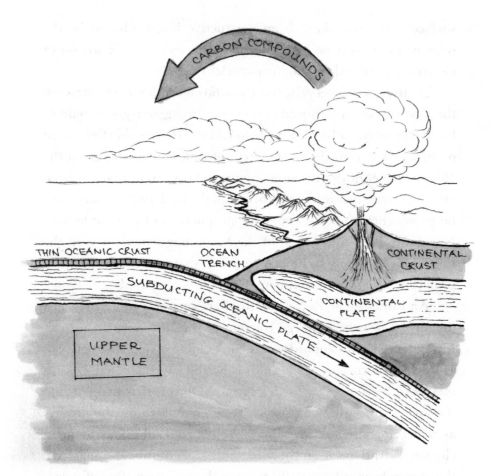

Over geological time, the carbon dioxide content of Earth's atmosphere is regulated by plate tectonics. CO_2 enters oceanic water directly or through rainfall to form carbonaceous rocks on the ocean floor. These are recycled into the atmosphere via subduction and volcanism.

Author, after USGS

layer of carbon that has fallen out of the atmosphere onto the ocean floor. The lubricating effect of seawater allows the subducting plate to descend a long way into the mantle, enriching its carbon content. Volcanic eruptions along the line of convergence then push that carbon back up into the atmosphere as carbon dioxide, from where it eventually falls again to the ocean floor. Given the fine balance of this complex process, there's little wonder that the additional atmospheric carbon dioxide from a century of fossil-fuel burning has a significant impact on global temperature.

HERE, IN THE PARCHED UPLANDS OF THE DARMARALAND wilderness, the atmosphere is thin, and the sunlight intense. It highlights a constant struggle taking place among the deep shadows of those giant granite blocks – the struggle of myriad species of African wildlife to survive. And it brings home a message to visitors like you and me. There is absolutely nothing typical about our planet. It is an extraordinary world, and caring for its atmosphere is something we could do better – much better. But, as an admirer of humankind's resilience, and an inveterate optimist, I'm willing to bet that like the tenacious flora and fauna of Namibia, we will fix it. A grassroots movement towards renewable energy was foreshadowed more than a decade ago by the late Hermann Scheer, a German politician and solar power advocate, and it's happening today. Hopefully, it will take effect soon enough to avert the peril of a runaway greenhouse effect like the one our next-door planet suffered some three billion years ago. With a surface temperature hovering around 470 °C and an atmosphere that drizzles sulphuric acid in its upper layers, Venus is not the kind of wilderness you'd ever want to visit. No matter how atypical it might be.

CHAPTER 2

★

THE TERMINATOR:
A USER'S GUIDE TO NIGHTFALL

Housed in its 17-storey-high dome, the Anglo-Australian Telescope boasts a dished mirror, 3.9 metres in diameter, that collects and focuses the light of faint celestial targets for detailed investigation by arrays of high-tech gadgetry. Its quarry ranges from nearby asteroids in the Solar System to the most distant objects detectable – exploding stars known as supernovae, and delinquent young galaxies known as quasars. And its vantage point on the Universe is a mountain-top called Siding Spring, located in New South Wales' Warrumbungle Range – a supremely apt Gamilaraay word meaning 'crooked mountains'. Siding Spring is 450 kilometres north of Canberra, and 350 kilometres north-west of Sydney. Its remoteness from major cities keeps its night skies as pollution-free as they were when the first humans watched the heavens from this place tens of thousands of years ago. Despite that, it's possible to discern the distant glow of Sydney on the horizon, along with other nearer centres such as Dubbo and Gilgandra.

When it was built in the early 1970s, the Anglo-Australian Telescope was one of the largest on the planet. Now, however, the world's astronomers have access to over a dozen telescopes with mirrors twice as big. Instruments of this size tend to be multi-national collaborations, but their physical locations concentrate in a handful of places on high mountain-tops not far from the western seaboards of continents. Here, exquisitely stable atmospheric conditions offer freedom from the blurring effects of turbulence, so Hawaii, the Canary Islands, south-western USA, northern Chile and South Africa are where the newest optical astronomy facilities are located. That's not to say Siding Spring's work is done: as we will see in chapter 15, clever technology combines with the site's dark skies to keep the observatory at the cutting edge of modern astronomy.

I was based at Siding Spring for more than two decades, using the Anglo-Australian Telescope and its smaller sibling – a wide-angle instrument known as the United Kingdom Schmidt Telescope. I got to know Siding Spring Mountain in all its guises, from crystal-clear sunsets that foretold perfect night-time conditions to gloomy fog-bound dawns, when the humidity had soared well above the dew-point.

The Earth's atmosphere plays a critically important role in the science that can be carried out in an observatory, of course. Its characteristics in terms of temperature, pressure, humidity, transparency and freedom from atmospheric turbulence and light pollution are the stock in trade of ground-based astronomers. We are, after all, gathering all our information through this fickle veil of life-giving gas.

Curiously, I discovered during my time at Siding Spring that the atmosphere can be as big an attraction as the sky itself when it comes to the simple pleasures of stargazing. And that is particularly true during that magical period when the dome of the sky is changing from daylight to darkness – or vice versa. Astronomers recognise that this twilight zone corresponds to our passage through the Earth's 'terminator' – a word that astronomers have used for centuries, but which has now been hijacked by the movie industry. In astronomy, it has serenely beautiful overtones; in the movies, it's a lot less serene.

So what is the terminator in this context? To understand it, you have to imagine yourself looking at a planet (or a satellite of a planet) from a vantage point in space. Moreover, you have to imagine said planet or satellite being illuminated by the Sun, as all Solar System objects are. The rest is easy, because the terminator is simply the line that divides the sunlit portion from the part in darkness.

For worlds with no atmosphere, like the planet Mercury or our own Moon, the terminator is a sharply defined boundary, startlingly abrupt as it delineates the change from darkness to light. But for a planet like Earth, with its blanket of air, the terminator blurs into a fuzzy line, with the illuminated side of the planet gradually merging into the darkness of the night side. That's because molecules in the atmosphere scatter the light of the Sun beyond the terminator's geometric boundary.

So let's now shift the focus back to our vantage point on the Earth's surface. As the planet spins on its axis, we are carried through the terminator twice in every 24 hours, experiencing either the gradual fall in illumination as the blue (or grey) of the daytime sky metamorphoses into the blackness of night, or, some hours later, the reverse. Just how many hours separate dusk and dawn depends on your latitude and the season of the year.

The period of twilight when the Earth's rotation carries us through the terminator is something most of us hardly notice. However, if you know how to look, it's a time when a rich assortment of atmospheric and astronomical phenomena manifest themselves. As you might expect, the sequences of events in the morning and evening twilight zones are perfectly symmetrical – that is, they are identical, but reversed in time. So, the following account applies equally to daybreak as to nightfall, except the order of everything is turned around. Since most of us experience dusk more often than dawn, we'll stick with the order of events at nightfall. Apart from anything else, it's a much more romantic time of the day.

IT MIGHT SEEM A BIT BASIC, BUT A GOOD STARTING POINT in understanding twilight phenomena is to ask why the daytime sky is bright. It's a question great thinkers throughout antiquity pondered, but it was not fully answered until the work of the English scientist, Lord Rayleigh, was published in 1871. Once again, it's the light-scattering effect of the atmosphere, whose subtleties we'll encounter in a couple of minutes. But anyone who has seen photos of the Moon's surface taken by Apollo astronauts in the late 1960s and early 70s will know that while they were taken during the lunar daytime, the sky itself is black. Without an atmosphere, the Moon has no means of scattering light, so the Sun's rays illuminate nothing until they hit the surface (or anything that might be standing on it – such as the odd astronaut). Actually, that's not quite true. Under certain circumstances, clouds of fine Moon dust are elevated by electrostatic forces in sufficient quantities to have been noticed by orbiting astronauts just before the Sun appeared over the rim of the Moon. But there's nowhere near enough of it to make the lunar sky bright.

So, back on Earth, if it weren't for the clouds that form in the atmosphere, our skies would always be blue. That colour comes from a particular aspect of the way sunlight interacts with air molecules and aerosols (dust particles or very fine droplets). Light is scattered in all directions by its interaction with these particles, but it turns out that its blue component is far more scattered than its red. As sunlight passes through the atmosphere, the blue light is extracted (which is why the Sun itself looks slightly yellow), but turns up again everywhere else in the sky. In fact, the violet light in the Sun's rainbow spectrum is scattered even more strongly than the blue, but it is also absorbed more strongly by the atmosphere – which is why our skies are a rich blue rather than a psychedelic violet.

When clouds are present, the blue is masked, of course, but the sky is still bright. The clouds themselves have a neutral shade ranging from brilliant white to a foreboding grey. The absence of colour in clouds is no accident: once again, sunlight is being scattered, but this time by water droplets that are much bigger than molecules and aerosols, and they don't follow Rayleigh's rule of blue supremacy. The droplets scatter all the colours of the rainbow equally, producing neutral white light (or grey light on dull days).

THE TWILIGHT PHENOMENA I WANT TO INTRODUCE YOU to are best seen when you have a clear horizon in all directions. Get away from buildings, trees, hills, mountains, dust storms, active volcanoes and other distractions. Mid-ocean is absolutely perfect, if you can manage it. But flat areas such as the high Karoo of South Africa, the Steppes of Asia or the deserts of the south-western United States are also ideal for this kind of viewing – not to mention the big sky country of inland Australia. Take a trip out to the Western Plains of New South Wales sometime, and make sure you visit Siding Spring Observatory in the process. But wherever you are, do choose a sunny evening for your twilight experience.

If you can, as the day comes to an end, set aside an hour or so to watch what happens as you cross the Earth's terminator. Sunset is the first piece of atmospheric enchantment to look for. As the Sun nears the horizon on its way down, you'll notice a distinct yellowing or even a slight reddening of the sky around it. That's because its light is travelling through a much greater thickness of atmosphere than when it is high in the sky, intensifying the removal of blue light, and even scattering some of the red light, too.

If there is dust or moisture in the atmosphere, perhaps with a few clouds blocking out the Sun, you can often see shafts of light radiating from its position in the sky. These are known as 'crepuscular rays' (evening rays), which, despite appearances, are actually parallel to each other. It is perspective that causes them to fan outwards from the Sun in that spectacular manner so beloved of landscape artists. Occasionally, faint crepuscular rays can be seen *after* sunset, in a sky that is completely clear. As before, their presence betrays the presence of clouds blocking chunks of the Sun's light, but these clouds are so far away as to be below the western horizon, and invisible to the viewer.

While the Sun is still low in the sky, turn your back to it, and have a look towards the east. Sometimes, you can see more crepuscular rays, now converging towards a point just below the eastern horizon that is directly opposite the Sun. This point is cleverly called the 'antisolar point', and the converging rays are … wait for it … 'antisolar crepuscular rays'. Even more surprising are antisolar crepuscular rays seen after the Sun has set, for their convergence point is now above the eastern horizon and slightly empty-looking – particularly in a cloudless sky. This is a fairly common occurrence at a mountain-top site like Siding Spring.

Just once, I have seen a crepuscular ray arching right across the sky from horizon to horizon, like a gigantic golden rainbow. It was on a humid Sydney summer evening, probably with a high aerosol content in the atmosphere. An amazing sight.

IF YOU HAVE A CLEAR WESTERN HORIZON, AND IT'S FREE from clouds at sunset, it is worth looking for the 'green flash'. Astronomers' friends and acquaintances often grumble that this is a figment of said astronomers' fevered imaginations, but the green flash is a real physical phenomenon. It can even be photographed. It's caused by sunlight being dispersed into an extremely short vertical rainbow spectrum because the Earth's atmosphere behaves like a prism. Most of the time, we simply don't notice it. At sunset, however, we see a diminishing proportion of the Sun's disc as it crosses the horizon from the first contact of its lower limb to its final disappearance – a process that lasts two to four minutes in North American latitudes. And right at the end, a fine sliver of brightness is left. Just occasionally, when the atmosphere is perfectly stable, this will turn bright green for a second or two before it disappears.

What's happening at this point is that the red and yellow components of the Sun's spectrum have now sunk below the horizon, due to the prismatic effect of the atmosphere. That leaves only its green and blue light in the final sliver. Our eyes are more sensitive to green light than blue, so we see an enhancement of green. It lasts only briefly, but it's quite unmistakeable when it occurs.

One problem with observing the green flash is that your eyes tend to be dazzled because you're constantly checking to see how near the Sun is to setting. No matter how much you try to avert your gaze, the radiant disc of our star demands your attention. For that reason, the green flash is best seen at dawn, when the first sliver of the Sun's disc emerges above the distant horizon. Of course, you have to know where to look, but that's not too difficult to work out from the brightening of the sky. The best green flashes I've seen have been at dawn.

GREEN FLASH OR NOT, ONCE THE SUN HAS DISAPPEARED below the horizon, turn again to the east to see one of the most poetic of all sunset phenomena. It's so commonplace that most of us don't even notice it, but once you know what you're actually looking at, you won't forget it. After sunset on a clear day, all along the eastern horizon you'll see a blue-grey band topped with a strip of pinkish purple light. As the Sun sinks further below the horizon, the blue-grey band broadens into a shallow arch whose apex is directly opposite the sunset. At the same time the pinkish glow becomes more prominent, separating the grey arch from the blue of the rest of the sky, sometimes with extraordinary brilliance.

What you're seeing here is the Earth's shadow cast on its atmosphere, rising majestically in the east as the Sun sets. As soon as the Sun has set, you are actually inside the shadow, and being carried eastwards away from its sharp upper edge by the Earth's rotation, which, in North American latitudes, ranges from about 800 to 1400 kilometres per hour. For this reason, the shadow soon becomes indistinct, and the grey arch and darkening sky gradually merge into a uniform blue-grey, as the pink glow disappears.

These delightful shadow effects have equally delightful names. The blue-grey arch is known as the *twilight wedge* (because of its three-dimensional shape in the atmosphere), while the pink glow is known more enigmatically as the *Belt of Venus*. Apparently it refers to the 'cestus' – a girdle or breast-band of the Greek goddess Aphrodite, aka the Roman goddess Venus. Its pinkish-purple colour comes from the fact that the atmosphere scatters the red-rich light of the setting Sun directly back towards the observer, and it mixes with the blue of the still-illuminated sky along the boundary of the Earth's shadow. It's a truly beautiful phenomenon that's commonly visible, but most people miss it.

AS THE SKY DARKENS, THE UNIVERSE STARTS TO REVEAL itself in all its splendour. Of course, the planets, stars and galaxies are there all the time, but they're hidden under the brilliance of the

WHAT YOU SEE:

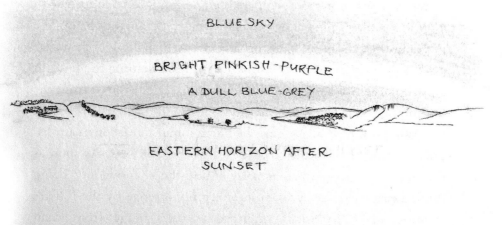

BLUE SKY

BRIGHT PINKISH-PURPLE

A DULL BLUE-GREY

EASTERN HORIZON AFTER SUNSET

WHAT'S HAPPENING:

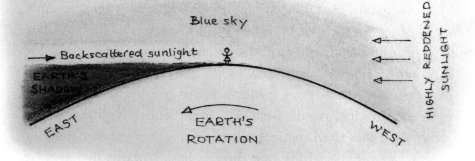

ATMOSPHERE

Blue sky

Backscattered sunlight

EARTH'S SHADOW

EAST

EARTH'S ROTATION

WEST

HIGHLY REDDENED SUNLIGHT

At sunset, reddened backscattered sunlight mixes with the blue of the sky to form a purple 'Belt of Venus' on the upper edge of Earth's shadow. As the planet rotates, the observer is carried into the shadow, and the Belt of Venus quickly becomes indistinct. The rotating Earth is viewed here from the north.

Author

daytime sky. One or two objects are brighter than the sky itself, so they can be seen in daylight. The Moon is one, of course, and at certain times, the planet Venus is another. Sometimes Venus is visible as a tiny speck of light while the Sun is still in the sky, when the planet is in, or near, a position known as 'greatest brilliance', which it occupies for a few days at a time, separated by intervals of a few months. The event's occurrence is dictated by the elaborate dance of our sister planet relative to Earth, and you can check the details via the internet (just put 'Venus greatest brilliance' into your search engine). Take care, though, because when this phenomenon occurs, the planet is relatively close to the Sun in the sky. Better not to risk using binoculars or a telescope for fear of accidentally beaming direct sunlight into your eye. That would be catastrophic.

Back to twilight, though. As the Sun dips further below the horizon, the brighter stars and planets become steadily more visible. You might be intrigued to know that astronomers define three stages of twilight, distinguished by the differing levels of sunlight still being scattered into the atmosphere. 'Civil twilight' lasts until the Sun is 6 degrees below the horizon, and while it holds sway, the sky is still quite bright. It's followed by 'nautical twilight', which takes the Sun to 12 degrees below the horizon, and then by 'astronomical twilight', which lasts until it's 18 degrees below. These definitions date from the late 19th century, and while they seem arbitrary, they were chosen so that by the end of astronomical twilight, there would be no scattered sunlight whatsoever in the sky. By then, the sky is 'officially' dark.

What happens next depends on the phase of the Moon, and whether you're skywatching from a place afflicted by light pollution, as most of our cities are. If the Moon is full, it lights up the sky with surprising intensity, allowing you to find your way around easily without artificial light. But if the Moon is a slender crescent, or absent from the evening sky altogether – and especially if you're well away from city lights – you might see the bright band of the Milky Way crossing the sky. This is our view through the thickness of our Galaxy's disc, and the Sun is just one of its 400 billion or so stars.

THERE'S ONE MORE PHENOMENON THAT NEEDS A CLEAR, moonless sky, completely free from light pollution, to be visible, but which is unmistakeable once seen. It is a faint pillar of light that projects upwards from the western horizon for half an hour or so after the end of astronomical twilight. The luminous pillar is called the '*zodiacal light*', and its axis lies along the ecliptic – the path of the Sun and planets through the sky. Look for it on spring evenings after dark, when the ecliptic stands more nearly vertically than at other times of the year in North American latitudes. Don't confuse it with the Milky Way, however, which is further along the horizon to the north.

It took a long time for scientists to figure out what caused this spectacle. One of my great scientific heroes, the Norwegian physicist Kristian Birkeland, barked completely up the wrong tree during the first decade or so of the 20th century by imagining it was due to an electromagnetic interaction between subatomic particles from the Sun, and the Earth's atmosphere. Birkeland had correctly identified this interaction as the source of the polar aurorae in the closing years of the 19th century, but in the case of the zodiacal light, his intuition let him down. A sojourn in Egypt to make detailed observations of the light was cut short by the First World War, and Birkeland made plans to return to his native Oslo. To avoid the hostilities (as well as the British, whose scientists had been scornful of his work on the aurora), he elected to travel from Cairo to Oslo via Tokyo. Not the most direct route, and, sadly, it was in Tokyo that he died from an overdose of sleeping medication, on 15 June 1917.

It wasn't long, however, before scientists worked out that the zodiacal light has a much more prosaic origin than electromagnetic interactions. Like the blue of the sky, it's a scattering phenomenon, with sunlight being scattered not by the Earth's atmosphere, but by particles of interplanetary dust in the disc of the Earth's orbit. These are large enough that they don't obey Lord Rayleigh's blue-biased rules of scattering, however, so – just like clouds in the sky – the zodiacal light is colourless.

THERE'S ONE FURTHER ASPECT OF THE ZODIACAL LIGHT that needs to be mentioned. This story goes back to the very early 1970s, when a young astronomy student in the United Kingdom began his PhD research on the topic at Imperial College, London. His task was to observe the rainbow spectrum of the zodiacal light over approximately a year from the high-altitude Observatorio del Teide in Tenerife. He was keen – but his career inconveniently lurched into music, at which he did tolerably well. Thoroughly neglected, his research quickly ground to a halt, until he picked it up again in the early 2000s, and graduated in August 2007 with his long-lost (and undoubtedly well-deserved) PhD. This now-eminent astronomer is better known as the lead guitarist of Queen – one Brian May. And it has to be said that there's nothing quite like a rock star for bringing street cred to the science of the Universe.

What is surprising, though, is how little interest the academic community had shown in the zodiacal light – so little, in fact, that more than three decades could elapse without Brian's research becoming totally outdated. But in the event, his timing was perfect. During recent years, the Solar System's faint dusty disc has received increasing attention from astronomers. It's now recognised as a fossil of the cloud of gas and dust from which the planets were born. It has much to tell us – far more, perhaps, than could ever have been guessed by a musically minded young astronomer stumbling forth on his research career.

CHAPTER 3

★

STARRING CITIZEN SCIENCE: RESEARCH BY THE PEOPLE

'The nerd side of me is just ecstatic!' You bet it is. This is a guy who has just discovered a four-planet solar system 600 light years from Earth. His name is Andrew Grey, and he's not an astronomer. At the time of this discovery he was a 26-year-old car mechanic from Darwin in far-northern Australia with a life-long interest in astronomy, whose persistence in trawling through a thousand or so light curves – star brightness graphs – had been rewarded big-time. And on live TV, to boot.

It was on *Stargazing Live* – a three-night TV blockbuster on Australia's national broadcaster, the ABC. The show sparked a frenzy of citizen science, and the challenge was to find the tell-tale signatures of planets orbiting distant stars in a mass of data from space. Known as exoplanets, these objects have been discovered in profusion over the past two decades, with more than 4000 known today. Very few have been seen directly, however, and most have revealed themselves by the subtle effects they have on their parent stars' light. The *Stargazing Live* data, for example, were newly downloaded from NASA's *Kepler* spacecraft, whose primary mission was to stare at 100 000 stars. Not just staring for the sake of it, but staring in the hope of recording minuscule dips in brightness that would reveal the passage of a planet across a star's disc. This so-called 'transit method' is today's gold standard for finding exoplanets, having netted most of those currently known. And Andrew found a star with not just one, but *four* transiting planets.

What is truly staggering about his find is how it compares with the discovery of the first exoplanet in 1995. Everything is different. The technology, the internet access, the level of popular interest – even the social environment that allows an enthusiastic amateur stargazer to participate in front-line scientific research, and 'to be published alongside people that went to university for years and years and for me, just a mechanic from Darwin, to have my name on it – I think it's pretty amazing'. So do I, Andrew, and hats off to you. (And, for the record, to the several dozen other Australians who were pipped at the post in discovering the same four-planet system during the show.)

STARGAZING LIVE MADE A BIG THING OF CITIZEN SCIENCE. But what is citizen science, and how does it work? The diction-ary definition is 'the collection and analysis of data relating to the natural world by members of the general public, typically as part of a collaborative project with professional scientists'. But citizen science means different things to different people, depending on what is being expected of the citizen. And there are similar terms, such as 'crowd sourcing'. Are they the same thing?

Well-known Swiss astrophysicist and citizen science propo-nent, Kevin Schawinski, is pretty definite about that. Referring to his best-known project, Galaxy Zoo, he says, 'we prefer to call this citizen science because it's a better description of what you're doing. You're a regular citizen but you're doing science. Crowd sourcing sounds a bit like you're just a member of the crowd – and you're not. You're our collaborator. You're pro-actively involved in the process of science by participating.' Citizen science is facilitated by modern technology, such as the internet and social media. But its history is much older than that. And it has been particularly fruitful in the field of astronomy.

The 19th century saw many examples of citizen scientists wielding their test tubes, microscopes or barometers in scientific pursuits, mostly upper- and middle-class men and women with the time, money and education to engage in such activities. Their results were often published (in journals such as the *English Mechanic and World of Science* – and even in prestigious scientific publications such as *Nature*), but it was really in astronomy that these pursuits were organised into what we'd recognise today as citizen science.

Amateur astronomers equipped with small telescopes have long been able to contribute measurements of the brightness of stars, for example. And for those stars whose brightness varies (unimaginatively known as variable stars), amateurs have provided a valuable service in monitoring their brightness fluctuations. This feeds into the professional field, where the variations can be interpreted in terms of the physical processes taking place inside the star. As long ago as 1911, variable star observers were organised into a cohort of citizen scientists by the American Association of Variable Star Observers. Nowhere was this symbiosis between amateur and professional astronomers more successful than in New Zealand, where a very small number of professionals relied on the wider resources of the nation's amateur astronomers.

Once again, it highlights the extraordinary value of the amateur community to astronomy, not just in New Zealand, but all over the world. That extends far beyond making a scientific contribution (although astronomy is one of the few sciences in which this is still possible). More significantly, it ensures that there is an accessible and widely available route by which anyone can become involved. The men and women of the amateur community also do much to popularise astronomy, organising lectures, discussion forums, star-parties, star-b-cues and a plethora of other events designed to introduce people to the delights (and pitfalls) of stargazing. For many of these enthusiastic individuals, the ability to carry out citizen science is just the icing on the cake.

CITIZEN SCIENCE OF A RATHER DIFFERENT KIND WAS pioneered by the SETI@home project, in which home computers were mustered into a distributed computing network. This allowed clever software to analyse huge quantities of data from large professional radio telescopes participating in SETI – a collective term for various well-directed searches for extraterrestrial intelligence.

In radio astronomy, the most prolific natural cosmic signature comes from cold hydrogen in space. This pervasive radiation has a characteristic wavelength, and has been studied since the 1950s. After a decade or so of using it in radio astronomy, SETI's proponents were arguing that the same wavelength might be used by galactic civilisations that wanted to signal to one another. Thus, the first observational SETI programs piggy-backed their monitoring systems onto the receivers being used for conventional radio-astronomy research – a formula that remains in use today. The spiky nature of an anticipated communications signal lends itself to computer detection, an aspect that led in May 1999 to the inauguration of SETI@home. The venture uses the spare capability of idle home computers to trawl through sets of data from piggy-backed monitoring systems, reporting results over the internet.

In fact, some radio telescopes do carry out dedicated SETI observations, rather than just piggy-backing onto conventional research programs. A recent high-profile example is the Break-through Listen project. The search for extraterrestrial intelligence has been successful in attracting philanthropic funding, no doubt because of its huge popular appeal, and a generous endowment has come from the Breakthrough Foundation, which is an initiative of a Russian investment tycoon by the name of Yuri Milner. Together with the late Stephen Hawking and several other scientific lumi-naries, Milner kicked off a multi-faceted exploration venture in 2015, of which the first component is Breakthrough Listen. It is, without question, the most ambitious SETI project to date. A US$100 million investment is being used to buy up to a quarter of the total observing time on two major radio telescopes – at the Parkes Observatory in Australia, and the Green Bank Observa-tory in West Virginia – along with technology enhancements that will also benefit conventional radio astronomy. SETI@home is an integral component of the data analysis.

Despite its longevity, SETI has so far failed to turn up any clear-cut candidates for extraterrestrial communication. Two events stand out: the famous 'Wow! signal' of 15 August 1977 (a short burst of radio radiation that prompted the eponymous com-ment scrawled on the print-out – and is still unexplained) and a signal from the direction of a star known to have an orbiting planet that Russian radio astronomers spotted in May 2015. That one eventually turned out to be coming from a secret military satellite. Ho hum.

WHILE SETI@HOME IS A TASK TO WHICH MACHINE INTEL-ligence is well suited, some large-scale astronomical observational programs are better suited to human pattern-recognition capabil-ities – and this is where citizen science really comes into its own. One such is the Galaxy Zoo project, founded in Oxford in 2007 by astrophysicists Kevin Schawinski (who we met a couple of pages ago) and Chris Lintott. It was modelled on an earlier NASA venture called Stardust@home, which required participants to visually scan 700 000 sets of images aimed at finding particles of interstellar dust collected and returned by a spacecraft called, yes … *Stardust*. The task required the judgment of trained volunteers, whose capabilities far outstripped the pattern-recognition software of the early 2000s.

Likewise, Galaxy Zoo is well suited to the capabilities of the human eye and brain, and is currently undertaking the biggest census of distant galaxies yet carried out. Galaxies are huge aggregations of billions of stars, of course, but they come in a wide variety of shapes, sizes, colours and other characteristics. While scientists understand broadly the origin and evolution of galaxies in their various categories, it's only by studying very large numbers of them that detailed characteristics can be established, and unusual outliers found. There are an estimated two trillion galaxies in the observable Universe, and a significant fraction have been imaged by the world's large telescopes, making their classification well suited to citizen science.

Galaxy Zoo has had several incarnations throughout its history. Highlights include discovering new categories of galaxies, such as the compact star-forming objects now known as 'green pea galaxies' (because that's what they look like). And it has also led to the identification of some very unusual celestial objects. Who could forget Hanny's Voorwerp (Hanny's Object), a rare light echo discovered by Dutch schoolteacher Hanny van Arkel? I'll discuss light echoes in more depth later on, but Hanny's Voorwerp is special, a light echo on a grand scale. A galaxy-sized cloud of gas has been ripped from a young galaxy by a passing interloper. But the disturbance has switched on something known as a quasar outburst in the young galaxy, and the black hole at its centre has begun consuming vast quantities of gas from its surroundings, while beaming intense ultra-violet radiation from its poles. The radiation has, in turn, excited the gas cloud, causing it to glow in a manner similar to a light echo. But by now, the quasar has switched off again, so all we see is an innocent-looking young galaxy with a wild-looking blob of glowing gas next to it – Hanny's Voorwerp.

Stories like that highlight the scientific value of the Galaxy Zoo – nothing quite like the Voorwerp had been seen before. A tenth anniversary conference at Oxford in 2017 celebrated the 125 million galaxy classifications and 60 peer-reviewed scientific papers that had been generated throughout the decade. The science even extended to psychological studies of biases in the perception of galaxy images. But there is also sociological value in the project. An online Galaxy Zoo forum spawned a genuine community spirit. As one prolific contributor, based in the Caribbean, put it, 'It was love at first sight when I started in Galaxy Zoo.' The venture is now accessed through a comprehensive web portal called the Zooniverse, which hosts almost 50 different citizen science projects covering a broad range of disciplines.

THUS IT WAS THAT CITIZEN SCIENCE WAS WRIT LARGE IN *Stargazing Live*. The events I described at the outset of this chapter took place back in 2017, when the show made its debut from Siding Spring Observatory in separate editions for British (BBC) and Australian (ABC) television. Hosted by megastar astronomer Brian Cox, it has segments on professional and amateur astronomy, as well as citizen science and record-breaking attempts for the greatest number of people doing sometimes obscure astronomical things.

As it is a live show, we had to broadcast before dawn in Australia for the BBC version to go to air in its usual mid-evening slot.

Early mornings are nothing new to astronomers, but dress rehearsals and studio make-up at 4.30 am were certainly a novelty. And in the end, the broadcasting powers-that-be were delighted with their million-plus audience numbers in both the United Kingdom and Australia. It highlighted the potential of social media to harness the enthusiasm of hundreds of thousands of people in the two countries. It was that more than anything that persuaded them that this was a venture worth repeating annually – in Australia at least.

In addition, the BBC version of the show highlighted a uniquely Australian citizen science project – the Desert Fireball Network's 'Fireballs in the Sky'. While the network itself consists of 50 automatic cameras constantly scanning the night skies of Western Australia and South Australia for bright meteors (shooting stars), its results are augmented with observations by citizen science participants. Together, these data allow fireballs to be tracked in three dimensions, raising the possibility that a meteorite might be recovered from the ground for scientific analysis. The project is led by Phil Bland of Curtin University, and the BBC interviewed Gretchen Benedix, the project's mineralogist/petrologist.

The two real-time citizen science projects featured in the 2017 *Stargazing Live* programmes addressed two of the highest profile issues in contemporary astronomy. The BBC highlighted the search for Planet Nine, a hypothetical world orbiting the Sun somewhere around 20 times further away than Neptune. More on that later. The project compared images gathered by the Australian National University's SkyMapper telescope at Siding Spring, with three separate photos of a given area of the sky, taken on different dates, being inspected to find slowly moving objects. A flurry of excitement ran through the *Stargazing Live* set when, during the final show in the series, a sequence of images was found that seemed to show an object in exactly the right part of the sky with the right amount of motion between its images. Sadly, it turned out that images of three different known asteroids had been captured, rather than three images of the same slowly moving object. Planet Nine thus remained elusive on *Stargazing Live* but, as we will see later, the hunt is still very much in progress.

A more successful outcome favoured the ABC's 2017 *Stargazing Live* project. Here, the citizen science challenge was that epic romp through data from NASA's *Kepler* spacecraft, which resulted in young Andrew Grey from Darwin hitting the exoplanet jackpot with his four planets. As the project's leader, Chris Lintott, pointed out, the discovery was scientifically important because there were only one or two other known solar systems where the planets were packed together so close to their parent star. And that might tell astronomers more about how planets form — one of the hot topics in current studies. For Chris, Andrew, and the millions of other ordinary people involved with citizen science projects around the world, this is a truly exciting and successful way to push back the frontiers of knowledge.

AND FOR ME, TOO, 2017'S *STARGAZING LIVE* BROUGHT SOME moments of unexpected excitement, including my formal introduction to asteroid no. 5691 live on the ABC version of the show, courtesy of the Las Cumbres Observatory's 2-metre telescope at Siding Spring. Even in such a large instrument, this perfectly ordinary — if not a little boring — main-belt asteroid looks like nothing more than a point of light. Since 2004, however, it has sported a name that seems somehow familiar — 5691 Fredwatson. Seeing that moving dot on the TV monitor definitely brought a sparkle to my eyes.

And finally, *Stargazing Live* opened those same eyes to something citizen science excels at, and has huge potential for the future of science generally. That's in the way it engages youngsters, giving them an opportunity both to learn and to contribute real knowledge via cleverly designed citizen science projects with appealing web interfaces. From palaeontology to stargazing, from spotting wildlife to tracking light pollution, there are projects for every interest. At a time when we need science more than ever to tackle environmental threats, this is a hopeful sign. Nothing less than the future of humanity is at stake. No pressure, kids.

CHAPTER 4

★

CATCH A FALLING STAR: METEORS, METEORITES AND SPACE DUST

Early one morning, well over a century ago, much of the population of the eastern United States was entranced, mystified and decidedly spooked by a dazzling display of shooting stars that all seemed to originate from the same point. They didn't have modern media to get the word out; nevertheless, they did pretty well. Most provincial newspapers carried accounts of this monumental meteor storm – perhaps the most active in recorded history, with 30 or 40 meteors raining down from the sky every second. It took place in the early morning of 13 November 1833, and was documented over the following three years by Denison Olmsted, Professor of Mathematics and Natural Philosophy at Yale College.

Olmsted's study was almost as monumental as the storm itself. His careful research was published in the *American Journal of Science and Arts*, and included many first-hand accounts, which were often very detailed. And it's Olmsted we have to thank for figuring out what was going on in this event. Because the myriad 'streams of light' seemed to diverge from a single point, he realised that objects were entering the Earth's atmosphere on parallel tracks. Perspective gave the impression that they originated in the constellation of Leo, which was high in the eastern sky before dawn. Olmsted interpreted this as being due to Earth passing through a dense cloud of particles, which themselves would have had a common motion through space.

We now know that this explanation is correct. The particles Olmsted surmised are specks of space dust or tiny stones, not much bigger than an orange pip, that hit the upper reaches of the Earth's atmosphere at high speed. They are instantly vaporised by the heat generated, and shine brilliantly for a few tenths of a second as they shoot across the sky. While they shine, they are known as *meteors*, a word originating in the 16th century to mean any atmospheric phenomenon, but now applied specifically to what are commonly called shooting stars. So – what is a meteor before it hits the atmosphere? Ah, there's a word for that, too: the slightly unfortunate *meteoroid*. And just to complete the trio, a *meteorite* is a meteor whose larger size allows it to survive its flight through the atmosphere, and reach the ground.

IN METEOR SHOWERS, THE PARTICLES ARE RELEASED FROM comets – 'dirty snowballs', a few kilometres across, which are icy remnants of the cloud of gas and dust from which the Solar System formed. They orbit the Sun in highly elongated paths, which often stretch well beyond the orbits of the planets. Comets become visible – and sometimes very prominent – when they reach perihelion, the point in their orbit when they are closest to our star. Here, the frozen gases that bind them together evaporate (or, more correctly, sublime) in the Sun's radiation, and can form bright tails of dust and gas.

Not surprisingly, comet orbits are littered with dusty debris thrown off during their visits to the inner Solar System. Clumps of debris share the orbit of the parent comet, and, during the Earth's annual tour around the Sun, our planet passes through a succession of these dust trails from a variety of different comets. The result is a well-established calendar of meteor showers, each of which appears to diverge from a point known as its 'radiant'. And each shower is named after the particular constellation containing the radiant – rather comically, with '-ids' stuck on the end. If the Earth's passage through the comet's orbit happens to coincide with a particularly dense clump of dust, then the meteor shower becomes a rare meteor storm – as happened in 1833, with the meteors known as the Leonids.

It was more than three decades after the 1833 meteor storm that the Leonids' parent comet was discovered. Around Christmas time in 1865, two astronomers by the names of Wilhelm Tempel and Horace Parnell Tuttle independently discovered a comet with an orbital period of 33 years. It reached its perihelion in 1866. And in Europe, in the November of that year, there was another impressive display of Leonids meteors. At two or three meteors per second, it was nowhere near as spectacular as the 1833 event, but still represented a remarkably rich meteor shower. The two following Novembers also saw good displays.

I'm sure you will have quickly done the arithmetic here, and noticed that the 1833 meteor storm must also have coincided with the perihelion passage of Comet Tempel–Tuttle, when the comet was relatively close to Earth. Clearly, these showers suggested that the dust in the comet's orbit was concentrated close to the comet itself. And sure enough, while 1899 and 1933 produced nothing out of the ordinary, in 1966 there was a meteor storm that, in the Americas, rivalled the 1833 display. Eyewitness accounts speak of 'a blizzard of meteors'.

By then, radar observations of meteors were possible, allowing astronomers to measure details of the Leonid dust particles. They turned out to be mostly lighter than average (around 0.01 grams) and burned up rather higher than average (at heights greater than 100 kilometres). The intersection of the orbit of the meteoroid stream with that of Earth results in an atmospheric entry speed of 72 kilometres per second, the fastest known. In the 1980s and 90s, further studies of the historical Leonid showers allowed astronomers to determine that most of the Leonid meteoroids trailed behind the comet, and slightly outside its orbit.

Investigations by a number of scientists, including my former colleagues at Siding Spring Observatory, David Asher and Robert McNaught, allowed them to map the dust density in the meteoroid cloud, which meant they could predict the Leonid meteor activity in 1999 and the first few years of the new millennium. Sure enough, there were good displays in 1999, 2001 and 2002, some of which I witnessed during my routine observing at Siding Spring. And Asher and McNaught's more recent work has refined the simplistic picture of a comet being accompanied by a few clouds of dust to discover exactly which perihelion passages of the comet have resulted in particular meteor displays. The 1833 meteor storm, for example, was caused not by that year's perihelion passage of Comet Tempel–Tuttle, but by the dust trail ejected by the comet during its previous visit in 1800.

Today, predictions of intense meteor storms are of interest not just because of their visual appeal, but because of the risk to all that essential high-tech infrastructure orbiting our planet. How would satellite communications have fared in 1833, I wonder?

MOVING ON FROM THE LEONIDS, IT'S EASY TO FIND ONLINE information about the other meteor showers that punctuate the year, along with their progenitor comets. The Orionids, for example, which are visible from anywhere in the world in late October,

originate from the dust trail of the most famous comet of all —
Comet Halley. And a favourite of mine, the Geminids, peak on
14 December — the birthday of the great 16th-century Danish
astronomer, Tycho Brahe (oh, and it's mine, too). They are highly
unusual in that their parent body is not a comet, but a dusty asteroid by the name of 3200 Phaethon.

Observing a meteor shower is something anyone can do. It's
interesting because, while a Leonid-like storm is unlikely, you
never quite know what's going to happen. And it's easy because
you don't need any technical equipment other than your own
patience. Shower meteors can flash across any part of the sky, so
there's no point in using binoculars or a telescope. The only thing
that marks them out as members of the shower is that they seem
to have come from the constellation after which they are named.
Gemini in December, for example.

So, the best observing accessories are warm clothing, a cup of
hot chocolate, and a comfortable chair from which to keep an eye
on the whole sky for an hour or so. A clear night with little or
no cloud is the main prerequisite, but if you can get away from
the light pollution of cities, so much the better. It also helps if you
check your calendar in advance, and select a meteor shower whose
appearance that particular year isn't going to be diluted by moon-
light. The full moon brightens the sky, and reduces the number of
meteors that can be seen. A lunar phase between new moon and
first quarter is best to aim for, because the Moon will set during the
first half of the night.

Which leads me to the one other requirement for good
shower-watching I need to mention. Unfortunately, it's rather less
cheering than warm clothes and hot chocolate. It's that you need
to be out and about in the small hours of the morning to catch
the shower. This is because the leading hemisphere of Earth — the
one running into the dust clouds of the meteoroid stream — is the

hemisphere you're in after midnight. Before midnight, your sky is facing backwards and you'll look in vain for shower meteors.

That's not to say there's no use in looking out for shooting stars in the early evening. There's still a good chance you'll see what are known as 'sporadic' meteors – ones that don't belong to a shower. They are simply particles of interplanetary dust in the plane of the Solar System, being tidily swept up by the atmosphere of our planet. They are the leftovers of planet formation, and can come from any direction in the sky – which is what distinguishes a sporadic from a shower meteor. Scientists estimate that the sum total of meteoritic material hitting the Earth's atmosphere every day is in the region of 50 tonnes, and possibly much more. That represents at least a billion individual meteors, which might seem hard to believe when you're standing under a stubbornly inactive sky waiting for something to happen.

As well as being briefly brilliant, meteors deliver a number of other odds and ends to the Earth's atmosphere as they burn up. There's a layer of sodium, for example, located in the upper atmosphere at a height of around 90 kilometres, which comes from meteors. Surprisingly, it's useful to astronomers, since it can be excited to glow with that familiar orange colour seen in sodium street lights. Upward-pointing high-power lasers are used to ener-gise 'artificial stars' in the sodium layer, which optical instruments can then lock onto. These are designed to remove the effects of atmospheric turbulence on astronomical observations – a tech-nique known as adaptive optics.

And burned-up meteors also leave behind their own trails of fine dust. The trails aggregate into high-altitude clouds that are occasionally made visible by the condensation of atmospheric ice onto the dust. They can be seen against the night sky when they are illuminated by the Sun long after sunset at ground level, and are known as noctilucent clouds – or 'frosted meteor smoke', as one poetic pundit nicely put it. These ethereal wisps of light occur predominantly in summertime at high northern and southern latitudes, although in recent years they have been seen nearer the equator and in greater numbers – perhaps as a result of climate change.

One other thing that the seasoned meteor-watcher might look out for is a *fireball*. In fact, you don't really need to look out at all, because if one comes along, you won't be able to miss it unless you're actually indoors with the curtains drawn and the doors shut. A fireball is a very bright meteor. According to the International Astronomical Union definition, it's one that is brighter than any of the planets. What that means, of course, is that it's brighter than the planet Venus, since Venus is the most luminous natural celestial object after the Sun and Moon. Perhaps one in half a million meteors satisfies this definition, and often it will be bright enough to light up the landscape like a flash of lightning. The green or reddish colouring sometimes seen is characteristic of oxygen atoms in the upper atmosphere being excited by the sudden input of energy, and then releasing that energy in the form of light.

If you do happen to see a fireball, it's worth listening out for a few minutes after the event. Occasionally, the sonic boom generated by its suicidal flight through the upper atmosphere is strong enough to reach the ground, travelling many tens of kilometres to be detectable as a dull thud.

WHEN THE METEOR ITSELF MAKES IT DOWN TO THE ground as a meteorite, it provides a valuable sample of extraterrestrial material – a free gift from the Universe. Once again, there are subtleties in terminology that are second nature to the specialists, but a bit baffling to the rest of us. A 'meteorite fall', for example, is one that has been tracked through the atmosphere before being recovered. Usually, it's tracked by visual observations, but a few have been located by automated systems such as the Desert Fireball Network. A meteorite that has not been spotted as it came through the atmosphere is known, fairly predictably, as a 'meteorite find'. *Finds* vastly outnumber *falls* in the world's scientific meteorite collections. And one other point to note is that meteorites are named after the place where they were recovered. That's nearly always somewhere on the Earth's surface, but a handful of meteorites have been identified on Mars and the Moon. Where but on the planet Mars could the Meridiani Planum meteorite have been found? (Yes, by NASA's *Opportunity* rover, in January 2005.)

Meteorites come in several different categories, but most of them are stony in composition, while about 5 per cent contain large amounts of iron. The stony meteorites are known as chondrites, and a large fraction of them are composed of small roundish particles that are remnants of the hot disc of dusty material in which the planets formed 4.6 billion years ago. The iron-rich meteorites, on the other hand, come from the cores of baby planets known as planetesimals – the building blocks of today's planets. Originally molten, the iron sank to the centres of these small worlds. The meteorites were subsequently knocked out of the solidified metal cores by collisions during the Solar System's early history, when planetesimals jostled together in the swirling disc of material surrounding the infant Sun.

Remarkably, iron meteorites have played a part in human history as well as planetary evolution. The ancient Egyptians were known to prize iron objects as long ago as 3400 BCE. In those days, iron would have been rarer than gold, because it wasn't until the sixth century BCE that iron smelting began there, as evidenced by archaeological studies. So where did that early iron come from? It came from the sky, the home of the gods – and analysis of Egyptian iron jewellery confirms its meteoritic origin, with high levels of nickel and cobalt. No wonder these items were regarded as precious – and none more so than a funerary dagger buried with the boy king Tutankhamun (1336–1327 BCE). Surmounted with a gold handle and sheath, it has an expertly crafted iron blade, whose composition closely matches that of a meteorite that fell a few hundred kilometres away on the Red Sea coast.

Perhaps the most special – and certainly the rarest – of all meteorites are those known to have been ejected from the Moon and Mars as a result of much larger asteroids hitting their surfaces long ago. More than 300 lunar meteorites are currently known, their identification hinging on their similarity to the samples of rock and soil recovered by Apollo astronauts. Analysis of their surfaces shows that most were ejected from the Moon within the past 100 000 years. And about 220 meteorites are known to have come from Mars. Once again, their Martian origin is deduced from chemical similarities with the atmosphere and rocks of Mars, as measured by robotic spacecraft.

The Martian meteorites are subdivided into groups with differing compositions, suggesting that they came from different locations on Mars. They are called shergottites, nakhlites and chassignites – names that come from the location on Earth where the first example of each class was found, in India, Egypt and France respectively. In fact, the majority of Martian meteorites are shergottites.

All these objects have been extremely well studied, particularly a 15-centimetre-long specimen by the name of ALH84001 – or more commonly, the Allan Hills Meteorite, named after the part of Antarctica in which it was found in 1984. Formerly regarded as a shergottite, it's now classified in a small group of its own. Famously, it contains tiny structures resembling terrestrial bacteria, which some scientists in the 1990s interpreted as fossilised Martian life-forms. Although the rock comprising

ALH84001 was formed around four billion years ago when Mars was warm and wet (making ALH84001 one of the oldest known Martian meteorites), most scientists today regard the identification of fossils as speculative at most, preferring a purely chemical origin for the structures. Nevertheless, the continuing interest in ALH84001 is a bonus for the science of astrobiology, providing a useful case study.

FINALLY, WHAT DO YOU CALL A FIREBALL THAT IS INCREDibly bright and breaks up in the atmosphere? Ah, that's a *'bolide'*, or if it's even bigger and brighter, a *'superbolide'*. While these terms sound a lot like hyperbole, they do come with technically defined intensities that needn't concern us here. But it's at this level where we begin transitioning into the realm of asteroid impacts and their effects – which is a whole other story.

Right on the transition is a recent event that hit the global headlines. It was the most significant impact of an extraterrestrial body since the Tunguska superbolide of June 1908, in which 2000 square kilometres of Siberian forest were flattened by an exploding object – a small asteroid or comet – some 5 kilometres above the ground. Coincidentally, the recent event also took place over Russian territory, at wintry Chelyabinsk, in the Ural Mountains.

On the morning of 15 February 2013 at sunrise, the skies over the district lit up with a brilliance 30 times greater than the Sun itself, as the streaking superbolide detonated above the city. Unlike the Tunguska event, which no-one seems to have witnessed, Chelyabinsk saw it all. The city abounded with security cameras and vehicle dash cameras, which provided an amazingly complete record of the incoming fireball and its dramatic explosion. A few people who were outdoors reported skin burns from the intense radiation. But the flash of light illuminated the snowy landscape without a sound, bringing those indoors to their windows to see what was happening.

Then, 88 seconds later, the shock wave arrived. Doors and windows blew in, complete with their frames; free-standing walls were demolished, and the roof of a warehouse collapsed. Some 1500 people had to seek medical attention – mostly with cuts from broken glass. There are stories of heroism, like that of teacher Yulia Karbysheva, who instructed her students to duck under their desks after the flash, but sustained serious injuries herself from flying glass. Some 100 000 home-owners were affected, and everyone struggled during the following days to keep warm in windowless buildings when outdoor temperatures were below −15 °C. But, mercifully, no-one died.

Soon after the event, people located meteoritic fragments to the south and west of the city, and discovered a large hole in the 70-centimetre-thick ice of Lake Chebarkul some 70 kilometres away. And, over the ensuing months, scientists gathered all the available information from orbiting spacecraft, dashboard cameras, security cameras, damage reports, seismometers, and fragments of the meteorite. The largest of these, recovered from the muddy bed of Lake Chebarkul on 16 October, weighed in at 650 kilograms.

By November 2013, the verdict was in. Two internationally renowned journals, *Science* and *Nature*, published the details. A 20-metre body weighing around 10 000 tonnes had caused the Chelyabinsk event by entering Earth's atmosphere at a speed of 19 kilometres per second. While its dimensions place it on the cusp of being an asteroid rather than a meteorite, it had eluded the world's asteroid-detection cameras because of its small size and its incoming direction – which was straight out of the Sun. It reached its peak brightness at an altitude of 29.5 kilometres, but exploded a few kilometres lower with an energy of some 500 kilotons of

TNT (equivalent to 30 Hiroshima blasts). Seismographs recorded a magnitude 2.7 tremor from the shock wave. And the explosion produced the largest atmospheric infrasound signal ever recorded, detected by 20 nuclear weapons monitoring stations, including one in Antarctica. Travelling at least twice around Earth, the infrasound waves took a full day to subside.

Where did the Chelyabinsk superbolide come from? Examination of the meteoritic debris shows it to have been an ordinary stony chondrite that was once part of a larger asteroid. And its trajectory could be accurately mapped after analysing all that camera footage, revealing an elongated orbit around the Sun. Its furthest point (aphelion) was in the main asteroid belt between the orbits of Mars and Jupiter, while its perihelion was, not unexpectedly, within the orbit of Earth. Intriguingly, there are similarities between the superbolide's orbit and that of a known Earth-crossing asteroid by the name of 1999 NC_{43}. It is thought that this asteroid itself suffered an impact a million or so years ago, creating an accompanying clump of rubble, of which the Chelyabinsk superbolide might have been a sample. Don't say this too loudly, but that might mean more are on the way.

DOES THIS MEAN WE SHOULD WORRY? IN FACT, NO CASES of death by meteor or asteroid impact have been recorded over the past 500 years. In bookending this chapter with the two most spectacular examples of celestial fireworks in recent history, I've tried to encompass the full range of what might be called 'normal' impact phenomena, from brilliant yet harmless cascades of milligram-sized dust particles to a decidedly dangerous object weighing thousands of tonnes.

Beyond that, though, things do get more hazardous. We know that impacts have significantly modified our planet's history, a theory pioneered in the late 1970s by two former colleagues of mine at the Royal Observatory, Edinburgh – Victor Clube and Bill Napier. Hot on the heels of their work came the realisation that the demise of the dinosaurs was probably the result of an impact 66 million years ago by an asteroid 15 kilometres in diameter, at a place now called Chicxulub in the Gulf of Mexico. But in the four decades since then, we have made enormous strides in understanding the Earth's environment. Today, the probabilities of objects of any given size hitting Earth are well-known. A Chelyabinsk-sized impactor might be expected somewhere in the world every 60 years; a Tunguska-sized one every thousand. And a Chicxulub-sized object will hit our planet roughly every 100 million years.

Statistics don't tell the whole story, however. On 18 December 2018, less than six years after the Chelyabinsk superbolide, a rather smaller object – about 12 metres wide – created an airburst fireball over the remote Kamchatka peninsula in eastern Russia. With a released energy of 173 kilotons of TNT, the event went unseen by human eyes, but was picked up by infrasound detectors and imagery from two unrelated research satellites. Once again, had there been any inhabited area beneath the impact site, windows would have been broken. Statistically, this is an event you'd expect to occur every 20 to 40 years, and the fact that it happened so soon after the Chelyabinsk impact highlights the stochastic nature of such phenomena. The curious coincidence of the three largest recorded meteor events since 1900 all occurring over Russia is accounted for by its size. Russia is by far the world's largest country by land area.

WE ARE NOW EQUIPPED WITH BATTERIES OF AUTOMATED telescopes searching for potentially hazardous asteroids. And, with the forewarning that they provide, there's every prospect of taking counter-measures against any threatened impact. Fortunately, the larger the object, the easier it is to find. It's estimated that 90 per cent of all hazardous asteroids bigger than 1 kilometre are already known, and search programs are now concentrating on objects down to 140 metres, of which 40 or so are discovered every month. Only a tiny fraction of asteroids are classified as potentially hazardous, and usually the level of threat from a particular object falls dramatically as its orbit becomes better characterised through ongoing observations. Small objects slipping through the net, like the Chelyabinsk superbolide, are rare occurrences, and will become rarer as the technology improves.

CHAPTER 5

★

RADIO SILENCE:
THE QUIETEST PLACE
IN THE WORLD

When people imagine what an astronomer does, they tend to envisage a geekish, white-coated person (usually a middle-aged, balding white male) peering through a long spindly telescope. Just, sort of, *looking* for things.

'Oh look, there's a nebula. Didn't see that yesterday. Better give it a name. How about 141244+031227B? What? Used that last week? Damn. Oh, wait – it's gone. Smudge on the lens. Where's the Windex?' And so on.

But that stereotype couldn't be more wrong. Pretty well all astronomy for the last century or so has been conducted with specific scientific questions in mind, as part of a directed global quest to understand the way the Universe works. Today, the telescope and its auxiliary equipment are at the absolute cutting edge of technology, which is one reason why governments fund the science of the stars – to drive the advancement of such technology. And, often, the telescope and its instruments are hundreds or even thousands of kilometres from the astronomer who's using them – who is much more likely to be young, female and from a more diverse background than they would have been in past generations. I'm honoured to count many young female astronomers among my colleagues, and am full of admiration for what they do. Lisa Harvey-Smith, for example, who is not only a distinguished professor of astronomy and science communication, but also the Australian government's first Women in STEM Ambassador. And Wiradjuri woman Kirsten Banks, whose work on Indigenous Australian astronomy complements her scientific studies.

Moreover, astronomers today make use of every variety of messenger that brings information from distant celestial objects to us. Subatomic particles and gravitational ripples in space-time are the most recent additions to the list of emissaries. Most of what we know about the Universe has come from electromagnetic radiation, however – the all-pervasive spectrum of vibrating electric and magnetic fields that conveys everything from gamma rays to radio waves, depending on the frequency of the vibrations.

Infrared astronomers, for example, look at heat radiation from space. Radio astronomers look at a whole bunch of natural radio emissions, X-ray astronomers investigate natural X-rays from very high energy sources, while optical astronomers use the ordinary visible light we're all familiar with. And just to answer an often-asked question, all these different types of astronomical observation are of equal importance – including the old-fashioned visible-light kind. In fact, by virtue of its central position in the overall spectrum of electromagnetic radiation, visible light provides a critical link between long and short wavelength observations.

Exploring the Universe with such a variety of wavebands allows scientists to build up a picture of what's going on out in space over a huge range of physical conditions – all of which give rise to differing emission processes. The hot surfaces of stars, for example, blaze with visible light – and, depending on their temperature, with ultra-violet or infrared radiation, too. On the other hand, cold molecules in space are efficient emitters of sub-millimetre radio waves. So, as you can imagine, having just one type of observation would be a bit like having just one piece in a jigsaw puzzle. Fortunately, that is far from the current situation.

MOST OF MY CAREER IN ASTRONOMY HAS BEEN INVOLVED with large optical telescopes (one of which, between you and me, did get its lenses cleaned with Windex). But some of the most exciting science today is being carried out in radio astronomy, in which Australia has a long history of excellence. This Australian expertise originated in radar research carried out throughout the Second World War at what was then known enigmatically as the Radiophysics Laboratory, located at Sydney University. Following the end of hostilities, the renamed Commonwealth Scientific and Industrial Research Organisation (CSIRO) Division of Radiophysics fragmented into a number of peacetime research groups, one of which was charged with investigating 'radio noise' from extraterrestrial sources.

At first, the group concentrated on radio emissions from the Sun, setting up an ingenious antenna on the cliff-top at Dover Heights, near South Head, Sydney Harbour. By 1947, however, the Dover Heights installation was being used to measure much more remote cosmic radio sources. And, incidentally, among the Australian scientists was the world's very first female radio astronomer, Ruby Payne-Scott (1912–1981). Ruby's story is one of extraordinary courage in the face of unjust sexual discrimination in science. Born and raised in New South Wales, she was educated at Sydney University, where she had a distinguished academic career. Her wartime work at the Radiophysics Laboratory led to significant postwar contributions in the new science of radio astronomy, together with the design of innovative new instrumentation and techniques. Forced to resign from CSIRO in 1951 because there was no such thing as maternity leave, Ruby raised her family and then returned to school-teaching, in which she had been briefly engaged before the war. Were she alive today, Ruby would be one of the megastars of STEM, and it's appropriate that in 2008, CSIRO established the Ruby Payne-Scott award to support staff returning from family-related leave.

It was the availability of huge quantities of wartime radar equipment, together with laboratories and staff to exploit them, that precipitated the rapid development of radio astronomy. At first, the purveyors of this new science were engineers, who were regarded with deep suspicion by their optical astronomy counterparts. For example, when Richard Woolley, then Director of the Commonwealth Observatory (today's Mount Stromlo Observatory), was asked in 1947 where he thought radio astronomy would be in ten years' time, he simply answered 'Forgotten'. Woolley was not known for his tact.

Eventually, though, the two wavebands were seen as complementary, and Australia became one of a handful of radio-astronomy centres of excellence throughout the world. The inauguration in 1961 of the iconic 64-metre-diameter radio telescope at Parkes in central west New South Wales enabled early studies of radio galaxies and their more exotic cousins, quasars, along with studies of the Milky Way and the rarefied gas between the stars. The telescope also made history with its role in NASA's Apollo program during the late 1960s and early 70s. Who could forget its (almost truthful) dramatisation in the 2000 movie *The Dish*?

I've always thought of the Parkes dish as one of the world's most picturesque observatories, comfortably settled as it is between rolling hills in the pastureland of the Goobang Valley, and looking for all the world as if it simply grew there. But today's most iconic Australian radio observatory has a very different demeanour. To see it, you'd have to journey to the remote inland of Western Australia, where red soil and hardy scrub dominate the wide-open landscape. Here, some 300 kilometres north-east of the coastal city of Geraldton, is the Murchison Radio-Astronomy Observatory, occupying a wide area of country whose traditional owners – the Wajarri Yamatji people – have watched the sky there for tens of thousands of years. Well-known Australian TV personality and Wajarri Yamatji elder, Ernie Dingo, was an enthusiastic visitor to the observatory in 2017. He commented on the new radio dishes with characteristic flair. 'This is wildflower country and they're like beautiful giant white wildflowers growing up out of the earth.' And so they are.

The observatory hosts a number of state-of-the-art telescope arrays. They include Australia's precursors to the Square Kilometre Array – the next major international project in radio astronomy, which will become the world's biggest telescope in the 2020s. As its name suggests, it will have a collecting area of a million square metres, and will comprise a very large number of individual antennas in Western Australia, linked to more in South Africa. To be exact, the South African array will really be a separate telescope, although it falls under the same umbrella organisation, known, unmemorably, as 'SKA'. The organisation's headquarters are at the United Kingdom's equivalent of the Parkes dish – the Jodrell Bank Radio Observatory, near Macclesfield in Cheshire.

In fact, both South Africa and Australia have been working on SKA pathfinder telescopes for the past decade or so – which are now operating observatories in their own right. In South Africa, the pathfinder has a rather splendid name that tells its own story – MeerKAT. So what's the story? Of course, you've seen images of meerkats, those cute little African mongoose-like critters, otherwise known as *Suricata suricatta*. But there's a clever twist to the name, since KAT used to be an acronym for the Karoo Array

Telescope, so-called after the proposed location of the instrument on the high Karoo some 450 kilometres north-east of Cape Town. Originally, it was planned to have 20 antennas, but when a generous budget increase by the South African government enlarged that to 64, it became, well … more KAT. Or, in Afrikaans, 'Meer-KAT'. Boom-tish.

It's a bit embarrassing to have to tell you that the equivalent at Murchison is just known as ASKAP – the Australian Square Kilometre Array Pathfinder. But, lest you should think our astronomers are completely devoid of imagination, I'm delighted to tell you that two of the survey projects taking place at Murchison are WALLABY and EMU, which, respectively, stand for Widefield ASKAP L-band Legacy All-sky Blind surveY, and Evolutionary Map of the Universe. Eat your heart out, MeerKAT.

While there is clearly some healthy competition between the African and Australian pathfinders of SKA, the two have complementary abilities. Yes, each has an array of similar-sized dishes – 36 antennas of 12 metres diameter each for ASKAP and 64 of 13.5 metres diameter each for MeerKAT. Both disgorge staggering amounts of digital data, too – not just from time to time, but every minute. However, their frequency ranges, while overlapping, are somewhat different. In general, South African astronomers are interested in higher frequency data than their Australian counterparts, and the specifications of their radio receivers reflect this.

THE MURCHISON OBSERVATORY BOASTS A NUMBER OF other new radio telescopes besides ASKAP, each built as much to trial new technologies for SKA as for scientific research. Most curious in appearance is the Murchison Wide-Field Array, which resembles a paddock full of large metal coat hangers arranged by someone obsessed with neatness. Its stationary antennas pick up low-frequency radio waves from the whole sky. Far less impressive-looking, it has to be said, is EDGES, which looks more like a large metal dining table – minus its dinner settings and accompanying chairs – than a radio telescope. But the US-operated EDGES has truly remarkable capabilities. While its acronym is a tad obscure (so much so that I won't even bother to relate it), its mission is a simple one. EDGES was built to detect the first stars to shine in the Universe. An ambitious task, and one with a touch of romance.

As you may be aware, science tells us that the Universe as we see it today started some 13.8 billion years ago in an explosive event that we have rather downplayed by naming it the Big Bang. For some hundreds of millions of years after the Big Bang, no stars shone, and astronomers can't help themselves but to refer to this period as the 'dark ages'. The quest to discover when the first stars flared into life and brought it to an end hinges on the ability of astronomers to look back in time as they look further out in space. An old trick, of course, which comes to us by courtesy of the finite speed of light. And radio waves.

But wait a minute – doesn't that mean that you'd have to look back by almost the whole age of the Universe to find the very first stars? And, at that distance, wouldn't individual stars be incredibly faint? The answer to both those questions is yes, but astronomers have another trick up their sleeves. The first stars didn't just shine with visible light. They also emitted copious amounts of ultra-violet radiation, which modified the cold hydrogen gas in which they were immersed. That gas was already emitting its own radio signal, but the change caused by the first stars imprinted a time-stamp on it, which should be detectable today. Crucially, the gas is spread over the whole sky, so you don't need to be looking in any particular direction. And EDGES can see everything above the horizon.

What this overgrown dining table has now revealed is that the first stars switched on only about 180 million years after the Big Bang. This is surprisingly early, but ties in with other measurements of youthful galaxies in the infant Universe. An even more surprising discovery is that the background gas was much colder than expected. A possible explanation for this could be that it was interacting with the mysterious dark matter, which is something that doesn't happen in today's Universe. As you'll discover in chapter 19, dark matter is by far the weightiest component of the Universe's mass, but reveals itself only by its gravitational attraction. The hint of a stronger interaction with normal matter in the early Universe could be a clue to dark matter's identity. It's a conclusion that remains to be independently confirmed at the time of writing, however, as does the exact epoch of the stars' switch-on.

The EDGES observation also highlights the singular value of the Murchison Observatory to radio astronomy. Just as optical astronomers require freedom from artificial light pollution to observe faint stars and galaxies, so radio astronomers need its equivalent, known as radio quietness. If you have a telescope capable of picking up the signal from a mobile phone at the distance of Pluto, it's not much good putting it near an urban centre where you're surrounded by a deafening cacophony of human-made radio signals.

The Murchison area fits the bill perfectly, with its remoteness, its low population density, and its freedom from radio communications traffic. Indeed, just as Siding Spring Observatory is protected by legislation from light pollution, so is Murchison shielded by a radio-quiet zone more than 500 kilometres in diameter, established by the Commonwealth and Western Australian governments. For this reason, although everyone at Murchison is proud of what they do, visitors are politely discouraged. Instead, they are urged to get online and seek out an engaging virtual tour of the site that gives a good impression of the facilities there. It's only when you realise that the incredibly weak signal detected by EDGES was found right in the middle of the VHF broadcast waveband that you begin to appreciate the true significance of Western Australia's most unsung natural asset. Its radio silence.

CHAPTER 6

★

THE OFF-PLANET ECONOMY: DOING BUSINESS IN SPACE

How would you like to buy a bit of the Moon? Or Mars? Or Venus or Mercury? Or anywhere else in the Universe, for that matter? Well, you can, from an organisation impressively named the Lunar Embassy. And the Lunar Embassy has impeccable credentials, having been founded by a former ventriloquist, actor and shoe salesman by the name of Dennis M Hope, of Gardnerville, Nevada. This gentleman claims to be the rightful owner of the Moon and various other celestial bodies. His assertion is predicated on the fact that in 1980, he wrote to the United Nations and the governments of the United States and Soviet Union asking them if they had any objections to his claim – and never got a reply. Since then, he has made a successful living by doling out portions of the surfaces of these worlds to clients who seem to take their deeds of lunar property very seriously. Personally, I'd be looking carefully at the fine print. Either way, Mr Hope has become a wealthy man.

The Lunar Embassy affair echoes another famous claim to ownership of a celestial body. That is the surprising assertion that the near-Earth asteroid, Eros, is the property of a Gregory W Nemitz of Twin Falls, Idaho. Back in 2000, Mr Nemitz lodged a claim of ownership with an organisation called the Archimedes Institute, which seems to have been created especially for the purpose. But when NASA successfully landed its NEAR-Shoemaker probe on Eros in February 2001, Nemitz sent the space agency a bill for $20 in parking fees. It was the first instalment of a 20 cents per year parking fee to run, well ... forever, which is how long the spacecraft is expected to remain there. NASA, of course, contested the account, and, after a lengthy legal process, the case was dismissed as Mr Nemitz could not prove that he actually owned the 33-kilometre-long asteroid.

These are perhaps the best-known instances of many wildly optimistic claims of extraterrestrial property rights that have been around for more than half a century. Comical though they might seem, they do have a serious side. For example, I'm sure Mr Nemitz would have been aware when he laid his claim that Eros is probably rather valuable. Back in 1999, it became one of the first asteroids to be recognised as having huge potential as a source of metals that are essential for the world's electronics industry, but frustratingly rare on Earth. It's estimated that Eros contains more platinum, gold, silver, zinc, aluminium and other metals than it would ever be possible to recover from the Earth's crust. A conservative projection in 1999 placed its value at US$20 trillion. Of course, if you dumped all those resources lock, stock and barrel into the world's metals market, their value would tumble, but the fact that they are so inaccessible suggests that is unlikely – for now, at least.

I THINK MOST PEOPLE REALISE TODAY THAT COMMERCIAL enterprise involving space is big business. With an annual global turnover approaching US$400 billion, it is a vastly different world from the one I was involved with in the 1960s. In those days, space was largely the province of two superpower governments: those of the Soviet Union and the United States. Its primary utilisation was military, with scientific exploitation clinging to its coat-tails as the poor relation – not that we didn't achieve a lot. And, of course, the driver in the human exploration of space was Cold War rivalry, with the Apollo lunar program representing the culmination of those early endeavours.

Today, space plays a part in almost every facet of human activity, with the commercial sector underwriting the lion's share of the cost. Communications, broadcasting, navigation, agriculture, weather forecasting, climate monitoring, resources and land-use management – the list is almost endless. And there is a rapidly increasing number of players in the market, mustered by an increasing number of national space agencies. As of mid-2019, no fewer than 72 space agencies were in operation, the most recent – surprisingly – being Australia's, which only came into existence on 1 July 2018.

Commercial investment in space received a major boost back in 2010, when the Obama Administration cancelled NASA's over-budget Constellation program, a three-stage venture designed to facilitate human occupancy of the International Space Station (ISS), and expedite travel to the Moon, and eventually, to Mars. Despite utilising technology originally developed for the Apollo and Space Shuttle programs, Constellation was ultimately deemed unsustainable without a significant increase in funding. In its place, a new vision was unveiled for the space agency that would allow it to concentrate on the cutting-edge technologies needed for future space exploration. The 'routine' work of servicing the ISS would be delegated to the commercial sector. That was already well placed to ramp up its development of necessary technologies, with several companies contracted to NASA for the provision of new launch vehicles and spacecraft.

SpaceX, for example – the space transport company founded by former PayPal whizz-kid and now avant-garde entrepreneur, Elon Musk – has developed its *Falcon* series of launch vehicles, together with the *Dragon* capsule that made history in May 2012, when it became the first privately operated spacecraft to deliver cargo to the ISS. Musk has continued to make the headlines with his bold efforts to improve sustainability in other areas than space, and his Tesla electric vehicles are rapidly becoming the new yardstick for car manufacturers. He also facilitated the installation of the world's biggest lithium-ion battery at a windfarm in South Australia in 2017, promising savings in energy costs that have now been realised. At the time of writing, Musk's *Crew Dragon* is on the brink of transferring astronauts to and from Earth-orbit, bringing to an end eight years of NASA reliance on Russian *Soyuz* vehicles for these taxi services. His biggest triumph in this area is that his first-stage rockets can be soft-landed for use in subsequent launches, reducing the cost of delivering material to low-Earth orbit from some US$20 000 per kilogram to around 10 per cent of that. His ambitions don't stop with *Falcon*, of course: a much larger space transport system, whose name curiously metamorphoses from time to time (most recently from the slightly suspect *Big Falcon Rocket* to *Starship*), is intended to ferry passengers to Mars in a venture that we will revisit in chapter 11.

There are several other US companies contracted to NASA to develop and operate medium-lift rockets, again with ISS cargo duties in mind. They include Northrop Grumman (which now operates the *Antares* rockets originally developed by Orbital Sciences) and the massive United Launch Alliance, which incorporates Boeing. Companies like these, together with other long-established space contractors like Airbus Defence and Space (which operates the European *Ariane* series of rockets), are familiar names in the annals of what might be called conventional

commercial spaceflight. Missions such as the launch and operation of unmanned communications and remote-sensing satellites,
specialised scientific satellites and, of course, military surveillance
satellites, have been the stock-in-trade of the commercial sector
for decades. But what new players like Musk and Jeff Bezos (the
Amazon founder whose Blue Origin company has also pioneered
reusable launch vehicles) have recognised is that serious money
might now be made from exploiting space more directly. These
entrepreneurs have sensed that the off-planet economy is an ambi-
tion whose time has come.

THE MOST VISIBLE SIGN OF THIS IS THE FLEDGLING OFF-
planet tourism industry. While we have heard talk of space tourism being just around the corner for almost two decades, it is still
not yet possible to buy a ticket and fly into space. That is mostly a
reflection of the sheer difficulty of implementing the technology to
achieve this safely. However, space tourism is not complete fiction.
Since the first space tourist, multimillionaire American engineer
Dennis Tito, took to the skies in a *Soyuz* space capsule in 2001,
there have been seven paying visitors to the ISS – one of whom
(software billionaire Charles Simonyi) went twice. These trips
were all brokered by a company called Space Adventures, which
used spare seats on Russian spacecraft to transfer their passengers
to and from the ISS.

Orbital space tourism of this kind is very expensive. None of
these passengers paid less than US$20 million for their trips, last-
ing between eight and 15 days each, and one is reputed to have
paid twice that amount. No wonder the Russian Space Agency,
Roscosmos, saw this as an effective way of boosting its flagging
fortunes while they had spare seats available. When NASA's
Space Shuttle program wound down, it effectively brought this

availability to an end, and the last paying customer flew in 2009. Space Adventures is continuing to market high-end space tourism, however, including a future lunar orbital mission with a ticket price in the order of US$150 million. Astonishingly, they claim to have a handful of takers.

Clearly, this kind of thing is never going to be mass-market tourism. So it has fallen to other visionaries to see the potential of a cheaper kind of space tourism, and, of these, none is more prominent (or flamboyant) than Sir Richard Branson. Through his company, Virgin Galactic, Branson is offering an experience of space for around only US$200 000 a ticket. And, while his first revenue-earning flight has yet to be made, he already has a waiting list of over 500 would-be passengers.

How will Branson provide spaceflights that are a hundred times cheaper than a *Soyuz* flight? The answer is that you don't go into orbit. You have a simple up-and-down flight profile that uses a rocket to kick you to a vertical speed of about a kilometre per second. A 'mother ship' carries the smaller space-plane to a height of about 16 kilometres, where it is released, and its rocket motor ignites. When that shuts down after a 90-second burn, your vehicle simply coasts on upwards until it begins falling back to Earth, from a maximum height of about 100 kilometres. While the coast phase is in progress, the craft and its occupants are in a state of weightlessness, which eventually comes to an end when aero-braking slows the craft to land like a glider on a conventional runway.

Branson's confidence in the space vehicles being developed for Virgin Galactic comes from the fact that his primary design contractor was a company called Scaled Composites (now also owned by Northrop Grumman), which was founded by another high-tech entrepreneur by the name of Burt Rutan. In 2004, Rutan's company won the US$10 million XPRIZE for exceeding a height of 100 kilometres in a privately operated piloted rocket plane twice within two weeks. That craft, eloquently named *SpaceShipOne*, was the model for the Virgin Galactic rocket planes. The Virgin test program has been much slower than Branson would have liked, however, with the first commercial flight having been 'about a year away' for nearly a decade now.

In April 2014, I had the good fortune to visit the recently completed Spaceport America near the town of Truth or Consequences (yes, that's really its name) in New Mexico. Everything seemed ready for Virgin Galactic to start operations from the king-size runway. It was not being used otherwise and the New Mexico government, which had underwritten the spaceport, was keen to see revenue-earning flights begin in order to start recouping its investment. But six months later, a tragic accident during a test flight destroyed the first Virgin space plane, the *VSS Enterprise*, causing the death of one of its crew. The ensuing National Transportation Safety Board investigation and the development of *Enterprise*'s replacement – the *VSS Unity* – cost Virgin Galactic more than three years. Early in 2019, however, the proving trials had reached the stage where the usual two test pilots were accompanied for the first time by a passenger – Virgin Galactic astronaut trainer Beth Moses, who was there to 'evaluate the passenger experience'. That's an encouraging sign, and perhaps by the time this book reaches the shelves, paying customers will be enjoying the experience first-hand, and relishing their view from space.

Suborbital tourism is not the exclusive province of Virgin Galactic, and a handful of other companies are undertaking comparable projects. Some have fallen by the wayside. XCOR Aerospace, for example, was developing its *Lynx* single-passenger rocket plane, which promised flights significantly cheaper than Virgin Galactic's. But the project was abandoned due to high development costs in 2016. Blue Origin has been much more successful, with Jeff Bezos's multi-faceted company preferring a conventional vertical lift-off for its tourist spacecraft over Branson's aircraft-type mother ship. What both Virgin Galactic and Blue Origin will give their space tourists is a view of the Earth's curved surface and its thin blue atmosphere from the blackness of space, together with about three minutes of weightlessness. It's an attractive prospect, and certain to be a life-changing experience for the participants when they're confronted face to face with the fragility of the biosphere. And, as the technology advances, it's likely to become cheaper and more widely available.

From a legal perspective, space tourism is currently in a similar situation to that of aviation a century or so ago. Safety is paramount — nothing would damage the infant venture more than the loss of a rocket plane and its passengers. On the other hand, over-regulation could stifle progress in tourism's Next Big Thing, so legislators have to face a delicate balance in exercising control. Currently, only a small number of countries have enacted laws and regulations allowing licensed operators to take paying passengers into space. The United States was first in 2004, and the United Kingdom's *Space Industry Act*, passed in 2018, includes a reference to tourism. Other nations will no doubt follow, since space tourism seems poised to become a lucrative business. One estimate of its future market value, published in 2018, arrives at a figure of US$1.27 billion by 2023.

IN TERMS OF REVENUE, HOWEVER, THAT SUM PALES INTO insignificance compared with the numbers being touted for perhaps the most audacious aspect of the off-planet economy — resource mining. Back in 2012, there was a flurry of commercial activity that led to two major companies setting up, declaring they intended to carry out space prospecting for rare minerals and metals on near-Earth asteroids, and then to mine them.

Planetary Resources (formerly Arkyd Astronautics) was founded by Eric Anderson (of Space Adventures), Peter Diamandis (XPRIZE founder) and Chris Lewicki (Mars Rover systems engineer and flight director). The slightly newer kid on the block, Deep Space Industries (DSI), was led by established space technologists David Gump and Rick Tumlinson. Both these companies were privately funded but, despite the deep pockets of their backers, they suffered financial problems that in late 2018 led to take-overs by other high-tech businesses: DSI by Bradford Space, and Planetary Resources by a blockchain company, ConsenSys, Inc.

A number of other smaller companies have since announced similar aspirations towards asteroid mining. While the current financial landscape is clearly volatile, it seems safe to assume that off-Earth resource extraction will eventually become a reality, and will follow similar milestones to those originally announced by Planetary Resources and DSI. They were to kick off by deploying fleets of small 'prospecting' spacecraft equipped with remote-sensing telescopes to scout out asteroids rich in the materials of interest. A by-product of this would be the discovery of near-Earth asteroids that might one day pose a collision threat to our planet. As both companies have declared a long-term goal of modifying the orbits of asteroids – primarily to make them more accessible for mining – there would be the evident benefit that the same technology could be used to avert a collision.

So, what's the motivation for mining asteroids? Primarily, as we noted in the case of Eros, it's because they are likely to be rich in metals such as nickel, platinum, palladium, osmium and rhodium; materials used in high-tech manufacture, but which are in relatively short supply on Earth. Typical of the figures quoted was an early Planetary Resources estimate of US$50 billion, as the value of the platinum alone from a 30-metre asteroid. And some types of asteroids are also rich in water, which increases their value further. It's present as ice bound in the asteroidal soil, or in hydrated clay minerals, and once extracted, can be dissociated into hydrogen and oxygen using solar-generated electricity, thereby producing rocket fuel.

The advantage offered by this dissociation technology is that the fuel doesn't need to be lifted from Earth, and both Planetary Resources and DSI envisaged setting up orbiting fuel depots for future space exploration. DSI hit the headlines early in 2013 with its assertion that a 40-metre-long asteroid known as 67943 Duende, which made a very close approach to Earth at the same time as chapter 4's Chelyabinsk meteorite (to which it was unrelated), might be worth US$195 billion in metals and recoverable water. (By the way, in case you're worried, despite its close approach in 2013, Duende poses no risk to Earth for at least a century.)

Another possibility DSI touted was a plan to harness another new technology, three-dimensional printing, to fabricate complex spacecraft components in orbit, avoiding the need to bring the raw materials to Earth altogether. This would be entirely robotic – as would be the mining operations themselves, with anything from a swarm of small spacecraft working together to very large units 'that look seriously industrial', as envisaged by Planetary Resources' Chris Lewicki.

Could asteroid mining work? From a practical point of view, there appear to be no show-stoppers, although the technology required is a long way from being available, and would be extremely expensive. For example, how do you extract material from the surface of an object that has barely enough gravity to hold itself together, and is rotating once every few minutes? And how do you attach machinery to a surface that may consist only of loosely bound rubble? The necessary technological developments could take decades to eventuate.

And the biggest question remains economic viability. Most

commentators agree that if minerals or metals were to be returned to Earth, the economic benefit would be minimal because the cost of vehicles that can re-enter the Earth's atmosphere is so high. Using the mined materials in space – especially water for rocket fuel – is more likely to be economically viable. But, for precious metals in particular, there is that other potential show-stopper mentioned at the outset. Even if the metals stay in space, abundant supplies could reduce prices below a level at which it is viable to extract them. Sounds like a catch-22 to me.

ONE THING THAT HAS MOVED ON IN AN ENCOURAGING way since plans for asteroid mining burst onto the scene in 2012 is the legal framework that would govern such thorny issues as who actually owns the resources. The underlying rules for this kind of activity are part of what is loosely termed space law, which is primarily embodied in the UN-ratified Outer Space Treaty of 1967 and its four additional conventions of 1968–79. These regulations were formulated at a time when the principal users of space were a couple of superpowers. By today's standards, they are incomplete and full of inconsistencies, and with private enterprise rapidly becoming the dominant force in space exploration and exploitation, it has become important to tie up the loose ends. And I guess it's fair to say that the progress that has been made with this has emerged from a pragmatic view of the Outer Space Treaty.

So, while Article II of the Treaty states that 'Outer space, including the Moon and other celestial bodies, is not subject to national appropriation by claim of sovereignty, by means of use or occupation, or by any other means', there is at least one precedent in international law that has given heart to would-be asteroid miners. That is the 382 kilograms of lunar rock and soil samples that the Apollo astronauts brought back to Earth, which no-one doubts are the property of the US government. If a government organisation can lay claim to material it has brought from space, why can't private enterprise?

It was exactly this kind of thinking that led, in November 2015, to the *US Commercial Space Launch Competitiveness Act*, which states in its Section 51303 that:

> A United States citizen engaged in commercial recovery of
> an asteroid resource or a space resource under this chapter
> shall be entitled to any asteroid resource or space resource
> obtained, including to possess, own, transport, use, and
> sell the asteroid resource or space resource obtained in
> accordance with applicable law, including the international
> obligations of the United States.

Critically, the Act confers ownership on resources only after they have been extracted. This circumvents any conflict with the provisions of the Outer Space Treaty itself, which says that you can't stake a claim to ownership of a celestial body.

And then, just under two years after the US Act came into law, the government of the small European nation of Luxembourg enacted similar legislation, with stirring words from the Deputy Prime Minister, Étienne Schneider:

> Luxembourg is the first adopter in Europe of a legal and
> regulatory framework recognising that space resources are
> capable of being owned by private companies. The Grand
> Duchy thus reinforces its position as a European hub for the
> exploration and use of space resources.

Unlike the US legislation, it includes the important exception that companies taking advantage of it do not need to be

Luxembourg-based. Despite that, several would-be asteroid mining companies now have offices in the small landlocked state.

I THINK THERE'S A VERY GOOD CHANCE THAT BY THE middle of the present century, many of the activities forecast in this chapter will have become reality. While they sound a lot like science fiction today, one only has to think of the way that the development of other sci-fi concepts such as mobile phones and satnav has been spurred on by commercial demand. If the demand is there, the technology will evolve.

By 2050, affordable space tourism is likely to have progressed beyond the suborbital to the fully orbital. That requires spacecraft to be able to achieve a horizontal velocity of 8 kilometres per second to remain in orbit, as opposed to the 1 kilometre per second upward shove needed to touch the edge of space at a height of 100 kilometres. Unconventional launch vehicles such as the *Skylon* hybrid jet/ rocket spaceplane being developed in the United Kingdom could, by then, have dramatically reduced the cost of getting into orbit. The ground-breaking SABRE (Synergetic Air Breathing Rocket Engine) propulsion units for this craft are currently under development by Reaction Engines Ltd, and received a significant boost in funding in 2018. Hints of where the tourists might stay when they get into space come from two prototype expandable hotel modules that are already in orbit, launched in 2006 and 2007 by Bigelow Aerospace, a company owned by hotel magnate Robert Bigelow. While they have never been inhabited, it's easy to imagine the stunning view from them. More recently, Bigelow Aerospace made the headlines with its Expandable Activity Module, which was deployed in 2016 as a compact extension to the ISS for long-term testing.

It's even possible there will be joy-flights to the Moon and Mars. We've already noted Space Adventures' plans for a Moon mission, and you can bet your life that the six-month transfer time to Mars will eventually be seen as a positive tourist experience, rather than a dangerously long exposure to hazards such as solar radiation. Back in 2013, pioneering space tourist, Dennis Tito, had some rather charming ideas for an 18-month fly-around-Mars trip through his Inspiration Mars Foundation. He suggested sending 'an older couple' on the trip, who had presumably already ironed out all their differences and would get on famously during the tour. While that seems like a plausible suggestion, the technical and financial challenges of such an escapade currently rank higher than the psychological ones, and the Foundation now seems to have gone very quiet.

By such exotic standards, the robotic mining of asteroids looks almost pedestrian. And it's my guess that at least some of the ideas being touted by today's off-planet resource visionaries will have come to fruition by 2050. Fuel depots in space, perhaps, circumventing the problem of tanking up a Mars lander with enough fuel to get off the planet's surface and back into an Earthward orbit. And there's the increasing likelihood of a permanently occupied base on the Moon, following NASA's accelerated lunar landing program mandated by the Trump Administration in 2019 – not to mention interest from other space agencies. India's and China's, for example. That could eventually lead to resource extraction from our satellite, most probably water for rocket fuel, but perhaps also Helium-3, an isotope rare on Earth that might eventually provide safe nuclear energy in a fusion reactor.

Many readers of these words will still be hale and hearty by mid-century, and perhaps participating in the off-planet economy. And who knows? Those optimistic clients of that renowned land agent of America's wild west, Mr Dennis M Hope, might get a first-hand look at their blocks of extraterrestrial real estate. One can only hope they'll be happy with them.

CHAPTER 7

MOONSTRUCK: WHERE DID OUR SATELLITE COME FROM?

You might think that after a lifetime of studying the Universe, I would have a few favourites among the planets, stars, nebulae and galaxies that litter the cosmos. Distant objects that have conjured up wonder and inspiration over the years. And, it's true, there are many that have. But, to be honest, the golden child in my list of cherished heavenly bodies is still our nearest neighbour – the Moon.

The reason for this fondness goes back to the earliest days of my interest in the sky. Using a marvellous old brass telescope borrowed from my school history teacher, I found great delight in exploring the lunar surface whenever there was a break in the rain-showers that scudded regularly across northern England. (Until they turned into snow-showers.) As targets for budding astronomers, the mountains, plains and craters of the Moon have always been hard to beat. And, for me, the Moon became a familiar celestial companion whose phases, eclipses, risings and settings always brought interest and satisfaction. For a while, it did get rather in the way as my work took me to fainter territory, brightening the sky when it was close to its full phase and hiding all the more distant stuff. In that regard, I've had a fortunate life, because my 'distant stuff' has encompassed everything from asteroids in the Solar System to quasars at the very limits of observability. And there was no need to worry – the other half of each month presented a darker vista, revealing everything an astrophysicist's heart could desire.

For centuries, astronomers took the Moon pretty much for granted. Scientists who studied our natural satellite were few and far between, since there really didn't seem to be much more that could be learned about it from our earthly vantage point. A dead world, with nothing happening. But a few astronomers – including the late great British science communicator, Sir Patrick Moore – observed occasional brief flashes of light on the Moon. Patrick christened these events TLPs, for transient lunar phenomena, when he co-authored a report for NASA in 1968. Most of them are probably caused by meteorite impacts, but low-level gas emission, due, perhaps, to residual volcanism, could also be responsible. There is now a systematic observing program for TLPs being carried out on a 1.2-metre-diameter telescope near Corinth in Greece, with support from the European Space Agency.

It was the lunar exploration of the *Apollo* era that really prompted science to sit up and take notice of the Moon. Altogether 382 kilograms of lunar rock and soil were returned by NASA's *Apollo* missions between 1969 and 1972. That huge collection of specimens permitted hands-on analysis of the physical structure of the Moon's surface for the first time. And since then, further samples have been recognised in the shape of the lunar meteorites mentioned in chapter 4. Who would have thought that almost half as much material again would arrive serendipitously on Earth as a free gift from the Moon? Of course, the drawback with the 190 kilograms or so of confirmed lunar meteorites is that their exact point of origin is unknown; they come from random locations on the Moon, with a sizeable fraction probably originating on the farside.

THE SCIENTIFIC STUDY OF SUCH A LARGE AMOUNT OF lunar material prompted a resurgence of interest in the Moon's origins, a question that had rather fallen by the wayside during the 20th century. 'Where did the Moon come from? Well, who cares?' But it was not always thus.

One of the earliest scientists to engage in a careful study of the problem was a chap with a familiar name – George Darwin, later Sir George. He was, indeed, the son of Charles, born in 1845, and clearly shared his father's interest in the origin of stuff. Rather than being a life scientist, however, George was Plumian Professor of Astronomy and Experimental Philosophy in the University of Cambridge, a position he attained in 1883. It was in Cambridge that he assembled his thoughts on the origin of the Moon in a scheme he called his 'fission theory'. The idea was that the young Earth was spinning much faster than it is now, and part of it 'blobbed off' to become the Moon. It postulated simple centrifugal force as the agency that created our satellite.

I well remember reading an account of this theory in an old encyclopaedia as a youngster. It was accompanied by a series of rather alarming sketches depicting the process. First came spinning Earth, flattened at its poles in response to the centrifugal force it experiences (which is actually the case, although to a much lesser extent than was depicted). Then it starts to grow a pear-shaped extension around its middle, taking on a decidedly maternal appearance until its bulge extends so far that it is attached to Earth only by a thin cord. When that breaks, of course, it frees the infant Moon to move out to its own orbit, while Earth wobbles around for a bit with a painful-looking scar on one side, only recovering to a sphere after some unspecified time has passed. Darwin's theory postulated that today's Pacific Ocean was the site of the Moon's birth, but a careful examination of the diagrams (which can be found on the internet today) suggests that the Moon somehow popped out of northern Australia.

Clearly whoever produced these diagrams was under the impression that Earth would behave like a lump of putty in space, which is fair enough, I suppose, given the general understanding of planetary physics in the early 20th century. I think the reaction of anyone coming across them today would be 'Oh my God – is that how it happened?' And the answer is no – the fission process would have been very different in reality. In order for it to work at all, the centrifugal force around the Earth's equator would have had to exceed the planet's surface gravity, and material would have spilled off into Earth orbit all the way around it. Not so much a blobbing as a dribbling. The debris would subsequently have coalesced by the process known as accretion, eventually forming the Moon.

The problem with Darwin's theory, and the reason it fell out of favour during the 20th century, is that the early Earth didn't have enough rotational energy ('angular momentum') to fall apart in this way. Because such energy is conserved, the pre-break-up Earth must have had the same rotational energy as the total in today's Earth–Moon system. Knowing that, we're able to calculate that it can't have been spinning any more rapidly than once every four hours. That's incredibly fast – but not fast enough. Only at a rotation speed of once every two hours would centrifugal force have been able to overcome gravity. And it seems Earth can never have spun so quickly.

WHAT OTHER POSSIBILITIES ARE THERE? THE SOLAR System's outer planets are thought to have gained at least some of their moons by capturing passing asteroids or icy objects from the distant Kuiper Belt beyond the orbit of Neptune. Gravitation holding sway, once again. Although today's Solar System is fairly neat and tidy, as befits its venerable age of 4.6 billion years, it was littered with the left-over debris of planet-building during the first half-billion or so years of its existence.

Most of this proto-planetary detritus was on a scale smaller than the Moon, which is quite big in the grand scheme of planetary satellites. In absolute size, it ranks fifth behind three of Jupiter's moons, and one of Saturn's. However, those objects all have parent bodies that are much larger than Earth. So, if you look at Solar System moons in relation to their parent planets, the Moon romps home as number one, with a hefty 1.2 per cent of the mass of Earth. (That's not true of some of the smaller bodies in the Solar System, however. For example, Charon, the largest moon of the dwarf planet Pluto, has about one-eighth of Pluto's mass.)

Even if you can find a wandering proto-planet as big as the Moon, the capture theory has a couple of difficulties that are, if anything, more serious than the rotation problem of Darwin's fission theory. The first is that Earth is too small to capture something that big. Its gravitational pull would be unlikely to hold onto the proto-Moon as it swung past. It's just possible that the young Earth might have had a thick and extensive atmosphere capable of slowing an incoming object to orbital speed by aerobraking, but that seems unlikely.

And then there's the oxygen isotope ratio issue to cope with. You probably have the same problem yourself. More seriously, you might be aware that chemical elements can occur in different isotopes, which are incarnations of the same element with differing numbers of neutrons in their atomic nuclei. (It's the number of *protons* in the nucleus that define which element it actually is. Protons are electrically charged, while neutrons aren't.) It turns out that the ratio of the stable isotopes of oxygen provides a characteristic fingerprint that is different for each object in the Solar System. But guess what? Earth and the Moon have identical oxygen isotope ratios, demonstrated by measurements of isotopes in the lunar material brought back by the Apollo astronauts. That fact seems like a killer for the idea of the Moon having formed separately from Earth, and then being captured.

IT'S ALSO A DIFFICULTY FOR WHAT IS UNDENIABLY THE most popular contemporary theory for the origin of the Moon – although it can be accommodated by clever modelling. This is the 'giant impact hypothesis', which postulates that within the first 100 million years or so of Earth's history, our planet suffered a glancing collision with another young planet about the size of Mars. The result was a spectacular plume of debris, which collected in orbit around Earth, and eventually accreted to form the Moon. So confident are the proponents of this theory that the hypothesised impacting body has been given a name. It's called Theia, after the mother of Selene, who is the goddess of the Moon in Greek mythology. Nicely done, whoever suggested that.

The hypothesised relative sizes of Earth and Theia (about two-to-one) are dictated once again by the angular momentum of the Earth–Moon system as it exists today. In this basic scenario, it turns out that most of the debris generated in the collision would come from the smaller body – Theia, the Mars-sized collider. But that's where the isotope problem rears its ugly head again. Since there is less than a 1 per cent chance that Earth and Theia would have had the same isotopic composition, you would expect the rocks of the Moon to have a different composition from the rocks of Earth. But, as we have seen, they are identical.

Some scientists have seen this as the death knell of the giant impact hypothesis, and have resorted to much more speculative ideas. One is the notion that the Moon was produced when a natural nuclear fission reactor in the early Earth exploded, driving off enough material to form the Moon. Such natural nuclear reactors are known to have existed in Gabon, Africa, about 1.7 billion years ago, so the idea is not totally off the wall. There is, however, no evidence of a giant explosion.

The majority of planetary scientists still see the impact as offering the most plausible scenario for the origin of the Moon. A number of investigators have looked at differing models that might account for the similar composition of Earth and the Moon, while also explaining the angular momentum of today's Earth–Moon system. One suggests that if the collision involved not a glancing blow from a Theia-sized impactor, but a steeper collision with a smaller object, and if the pre-collision Earth was spinning faster than has been supposed, the impact would raise Earth material into orbit, rather than material from the impactor. It also invokes the gravitational attraction of the Sun in slowing the Earth's spin to yield today's observed angular momentum.

Another theory dispenses with Earth and Theia altogether and instead imagines two 'super-Theias' coming together in a relatively slow collision – a head-on bingle at just a few kilometres per second. The resulting debris cloud would be a mixture of material from the two objects. Both Earth and the Moon would have formed from it, explaining their identical composition, while the slow collision speed would account for the current observed angular momentum.

But new research published in April 2019 by scientists in Japan and the United States may hold the key to the problem. They have looked at what would have happened if the collision with Theia occurred so early in Earth's history that its surface was still an ocean of molten magma, as it was for the first 50 million or so years. They postulate that Theia had, by then, solidified, being a much less massive object. Their modelling shows that under those circumstances, the plume of ejected material that became the Moon would have been composed mostly of terrestrial magma, rather than rocky debris from Theia. Introducing this temperature differential between Theia and the proto-Earth represents a major advance in our understanding of the Moon's origin.

WHILE PEOPLE WILL CONTINUE TO DEBATE THE DIFFERences between the various flavours of the giant impact hypothesis, the one common element is that the Moon formed from the orbiting debris the collision produced. A swirling cloud of material, gradually accreting into a spinning globe. So how, then, did today's situation arise, in which the Moon always turns the same face to Earth? And why does that familiar face have such different characteristics from those on the Moon's farside, as was observed very early in its exploration by orbiting spacecraft in the 1950s and 60s?

Starting with the first of those questions, the short answer is tidal friction. It means that the tides raised by the Moon on the oceans and continents of Earth, and those raised by Earth on the rocks of the Moon (yes, it's true!) impart a transfer of energy between the two. The Moon actually gains energy, and responds by moving slowly away from us, at a rate of 3.82 centimetres per year. But as well as pushing the Moon away from Earth, tidal friction slows down the rotation of both bodies. The Earth's gradual slow-down is one reason we have to insert leap seconds into our timekeeping every so often. The Moon, however, being a smaller object, long since attained the final stage in this process. It is said to be 'tidally locked', which means that it rotates on its axis in the same length of time that it revolves around our planet. In other words, it always points the same face towards us.

The answer to the second question – about the differing Moonscapes of the near and far sides – has a less certain answer. But what is certain is that the so-called 'lunar dichotomy' is real, and the two are very different. We're all accustomed to the near-side, of course. Even a casual glance at the full Moon with the unaided eye shows that its surface is blotchy, with greyish patches that form a face, or a kangaroo, or a rabbit, or whatever else you're deceived into seeing by the persuasive psychological phenomenon of pareidolia. If you have access to binoculars, you'll also notice that the brighter regions are mountainous, and pockmarked with craters. These are the lunar highlands, rising much higher than the grey plains, which are actually ancient lava flows. That doesn't stop us referring to them as *maria* – the Latin word for seas, which is what early sky-watchers thought they were.

In its overall demeanour, the lunar farside is much more like the mountainous zones of the nearside. Lots of craters, a couple of small, isolated *maria*, and one of the biggest dents in the entire Solar System. This is the South Pole–Aitken basin, the result of a violent impact very early in the Solar System's history. It's likely that the collision excavated material from deep in the Moon's rocky crust, making it a tempting target for geological exploration. Which is why the intrepid Chinese spacecraft *Chang'e 4* landed there in January 2019.

There's plenty of evidence that the Moon's crust is rather thicker on the farside than on the nearside. The lava flows that formed the nearside *maria* had no trouble seeping through the thin crust to fill in the large impact basins that were formed at around the same time. So why should this dichotomy exist? Once again, good old conservation of angular momentum comes to the rescue. It suggests that the new-born Moon orbited our planet at only about 10 per cent of its present distance – just a fraction further out than today's geostationary communications satellites. Being so close, the gravitational interplay between Earth and the Moon would have very quickly locked the Moon into its current mode of rotation, with the same face always turned to Earth. And, crucially, at that

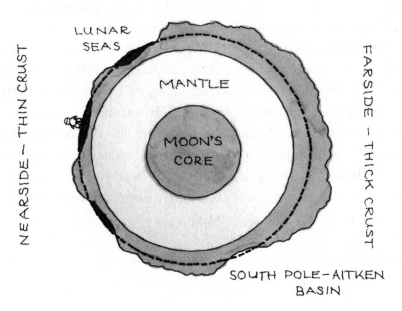

EARTH

384 400 km

LUNAR
SEAS

MANTLE

MOON'S
CORE

NEARSIDE – THIN CRUST

FARSIDE – THICK CRUST

SOUTH POLE–AITKEN
BASIN

(NOT TO SCALE)

Schematic cross-section of the Moon showing the thicker crust on the farside (highly exaggerated). Lava flows that pooled in large impact basins to form the nearside 'seas' failed to penetrate the farside crust. The cartoon astronaut stands near the first Apollo landing site.

Author, after Kenneth R. Lang

early stage in its history, Earth was still extremely hot. The Moon's nearside, therefore, was subjected to a high degree of radiant heat, which would have inhibited the formation of a thick crust. Cooler conditions on the farside allowed rock-forming elements such as calcium, aluminium and silicon to condense more rapidly.

Not all planetary scientists subscribe to this view of the thicker farside crust's origins, however. In 2011, a competing hypothesis suggested that the collision that originally produced the Moon also produced a second, smaller body, which eventually collided with the Moon's farside. The physics of this event are, we're told, 'consistent with the dimensions of the farside highlands'. As an inveterate admirer of the Moon's serenity and beauty, it saddens me to have to tell you that for want of a better name, this hypothesis has come to be known as … 'the big splat'.

PLANETARY
EXPLORATIONS

CHAPTER 8
TELESCOPE TROUBLES: ASTRONOMERS IN COURT

Most astronomers are fascinated by the history of their subject. Perhaps it's not surprising, since a common way of teaching astronomy is to trace its historical development, from superstition to reason, and from ignorance to ... well, a little less ignorance. And I confess I'm as susceptible to this fascination as anybody.

In 2009, the world's astronomical community celebrated the International Year of Astronomy (IYA) to mark the four-hundredth anniversary of Galileo first turning a telescope on the night sky. 'This was the beginning of modern instrumental astronomy, and a milestone in the history of evidence-based science', ran the International Astronomical Union's promotional blurb. Indeed, it was – but perhaps more unexpectedly, it was also the beginning of disputes concerning telescopes and what they might reveal. And such controversies continue, even today.

GALILEO'S CASE WAS CERTAINLY THE MOST EPOCH-MAKING as far as the course of science is concerned. The telescope he used was not his own invention, but had turned up in the hands of Dutch spectacle makers in the northern autumn of 1608. At least, that was when it appeared in the historical record; there's evidence that the idea of using combinations of lenses or curved mirrors to magnify distant objects had been around for much longer – even if they had never quite made it to reality. Chaucer mentioned such things in his *Canterbury Tales* in the late 1300s, for example. It was intense diplomatic activity that finally brought the 'far-seer' out of the woodwork, however, with Spain and the Netherlands locked in difficult negotiations to halt the war they had been fighting for the previous 40 years. An enterprising spectacle maker by the name of Hans Lipperhey arrived at the seat of Dutch government in The Hague, seeking a patent for this useful piece of military hardware. But within three weeks, two other individuals had filed counterclaims for the invention, and the result was that no patents were awarded. The cat was out of the bag, and word of the invention spread rapidly.

By May 1609, the news had reached our man – Galileo di Vincenzo Bonaiuti de' Galilei, the capable professor of mathematics in the University of Padua. His insight enabled him to formulate the optical prescription needed to make a telescope, and then perfect the lens-grinding process necessary to bring it to reality. In fact, he made at least four versions with successively higher magnifications, culminating in one that made distant objects appear 30 times larger. It was with this impressive instrument that he embarked on a spree of celestial discovery towards the end of 1609. Mountains on the Moon, congealed stars rather than congealed milk in the Milky Way, and most significantly, four satellites 'flying about the star [sic] Jupiter ... with wonderful swiftness', rocketed Galileo to international fame when he published his findings in a little book in March 1610.

Sidereus nuncius – or *Starry Messenger* – is a truly fascinating read if you can get hold of one of the many translations from the original Latin that are now available. And it is beautifully illustrated with Galileo's own depictions of the Moon, star clusters and the back and forth motion of Jupiter's moons. While the book did not explicitly support Copernicus's controversial theory that the Sun was at the centre of the Solar System (published by the great Polish astronomer nearly seven decades earlier), the moons of Jupiter clearly demonstrated that not everything revolved around Earth.

Towards the end of 1610, Galileo made another discovery. His telescope revealed that the planet Venus, which appears to the unaided eye only as a brilliant star in the morning or evening sky, actually displays phases like the Moon. With both its 'full' and 'new' phases occurring when Venus was close to the Sun in the sky, it had to be in orbit around the Sun, rather than Earth. Here was the seed of Copernicanism, which had already been planted in Galileo's mind more than a decade earlier. But it was a dangerous idea, at odds with the teaching of the Holy Roman Church. That all-powerful body held to the Aristotelian (or Ptolemaic) view that Earth is at the centre of the Universe, and everything moves around it. And support of Copernicus's view of the Solar System was one of the misdemeanours that had taken Giordano Bruno – 'the mad priest of the Sun' – to the stake in Rome's Campo de' Fiori on 17 February 1600.

In 1613, Galileo wrote a lengthy new book, *Istoria e Dimostrazioni intorno alle Macchie Solari*, usually known in English as *Letters on Sunspots*. Copiously illustrated with sketches of sunspots, explanatory drawings and, rather unexpectedly, diagrammatic predictions of the movements of Jupiter's moons, the book challenges the Aristotelian idea of flawless perfection in the Sun, and lays down the gauntlet of Copernicanism. It is particularly critical of the work of a Jesuit astronomer, Christoph Scheiner, whose observations had led him to interpret sunspots as clusters of small bodies orbiting the Sun, thereby preserving its Aristotelian perfection. After all, Galileo had discovered objects randomly circulating around Jupiter, so why not invoke objects randomly circulating around the Sun to explain the mysterious spots? Ah, retorted Galileo in his *Letters*, the movement of Jupiter's moons can be accurately predicted – hence the diagrams. The motion of the spots can't, and must therefore be flaws in the solar surface itself.

Stirring up Scheiner was probably a mistake. It pitted the Jesuit community against Galileo, who was already feeling the heat from another adversary, a Dominican friar and committed Aristotelian by the name of Tommaso Caccini. It was Friar Caccini who, in March 1615, lodged a formal complaint about Galileo's perceived impieties to the Holy Office, citing his *Letters* and other writings. By then, Galileo was firmly established in Florence, but he determined to travel to Rome to clear his name. However, his name was already before the Congregation for the Doctrine of the Faith – otherwise known as the Holy Roman Inquisition – which began its investigations towards the end of that year. A group of learned theologians, known as the Qualifiers, or Consultors, deliberated on the merits of a heliocentric (Sun-centred) model of the Solar System, and, on 24 February 1616, presented their report to the Inquisition.

They concluded unanimously that the idea of a static Sun is 'foolish and absurd in philosophy, and formally heretical since it explicitly contradicts in many places the sense of Holy Scripture'. Likewise, the proposal that Earth moves around the Sun was given short shrift. The next day, Pope Paul V convened a meeting of his cardinals, and instructed one Robert Bellarmine to communicate the outcome to Galileo. This Cardinal Bellarmine did so in a meeting at his residence on 26 February, a meeting that turned out to be crucial in Galileo's subsequent travails.

It's understood that Bellarmine himself was not opposed to Copernicanism, so long as it was used merely as a device for calculation, and not as a representation of physical reality. But at this point in the narrative, things become fuzzy, as there are two versions of the meeting's outcome. One is that Bellarmine instructed Galileo not to hold or to defend the Copernican claim of the Earth's motion, warning him that if he failed to acquiesce, he would be imprisoned. This was Galileo's impression, and he asked Bellarmine to confirm it with a letter, to squash rumours of his trial and condemnation – which Bellarmine did. It was a reasonably satisfactory outcome for Galileo, who, duly admonished, returned to his studies, hampered by the knowledge that he could not publish what he knew to be true, but would be spared the baleful glare of the Inquisition.

However, the other version includes a 'Special Injunction' by the Commissary General of the Holy Office, which ordered Galileo to abandon the Copernican model altogether, stating that he was 'henceforth not to hold, teach, or defend it in any way whatever, either orally or in writing; otherwise the Holy Office would start proceedings against him'. Although Galileo's signature is missing from the document, it goes on to state that he agreed to these terms, and undertook to obey them. Thus were the seeds of Galileo's eventual downfall sown, for this injunction inexplicably vanished from the record for 16 years, surfacing (to Galileo's surprise) on the eve of his trial in 1633.

IT IS ARGUABLE THAT GALILEO BROUGHT ABOUT HIS eventual trial himself by misjudging attitudes within the Church. On 6 August 1623, an old champion of his work by the name of Maffeo Barberini became Pope Urban VIII, and, in a series of audiences the following spring, Galileo discussed Copernicanism with a freedom that suggested the hypothesis might have found some acceptance. But that acceptance probably hinged on regarding the Copernican model merely as a tool for calculation rather than a representation of physical reality, in much the same way as Bellarmine had viewed it, thus avoiding contradicting the Scriptures. One of today's foremost Galileo scholars, Maurice Finocchiaro of the University of Nevada, has suggested that this differed from Galileo's view of what constitutes a hypothesis, which probably aligned more with the modern view – that it is an as-yet-unproven representation of a physical reality.

Encouraged by his meetings with Urban VIII, Galileo set about his next major task, a book that would compare the Earth-centred and Sun-centred models of the Solar System with particular regard to the phenomenon of tides on Earth. In fact, he was barking up the wrong tree, since the occurrence of tides doesn't prove the motion of Earth. Nevertheless, other arguments in his book strongly supported the Sun-centred Copernican model – the phases of Venus, for example – and in its overall tenor, the book advocated the dangerously heretical Copernican view.

Galileo used the literary device of dialogue to present his arguments. Like his *Letters on Sunspots*, the book was written in the common language – Italian, rather than Latin – to give it a broader appeal. Its original title when he submitted it to the Church authorities for approval was *Dialogue on the Ebb and Flow of the Sea*. But this hinted that the real physical phenomenon of tidal motion was a consequence of the Copernican hypothesis, and so he was instructed to change it, along with some alterations the Pope had suggested to the text. So, early in 1632 – and with the imprimatur of the Inquisition – the book appeared under the title *Dialogue by Galileo Galilei on the two Chief World Systems, Ptolemaic and Copernican*.

For his protagonists, Galileo had invented three individuals, undoubtedly modelled on friends and enemies in his circle. Salviati was effectively Galileo's mouthpiece, arguing for the Copernican position. Sagredo was an intelligent and impartial observer, supposedly from Venice, where tides are rather important. And Simplicio was an incompetent Aristotelian, clinging to the naïve view that Earth is stationary, and at the centre of the Solar System.

In the preface to the Dialogue, Galileo states that his choice of the name Simplicio was in homage to Simplicius of Cilicia, a distinguished 6th-century exponent of Aristotle's views. But, of course, the name is also close to the word 'simpleton' in many European languages, including Italian. And to make matters worse for Galileo, the Aristotelian arguments that Urban VIII had asked him to include turned up in the words of the idiotic Simplicio. Bad move.

Predictably miffed, it was Urban VIII himself who referred the book to a special commission a few months after its publication, by which time it had already done rather well in the bookshops. Further sales were immediately prohibited. From there, Galileo's path to the Inquisition was inevitable. A tribunal of ten cardinals made up the jury for his trial, which took place in Rome from April to June 1633. Galileo was interrogated, and confronted with the Special Injunction. In regard to that, Maurice Finocchiaro speculates that Galileo was actually framed. Perhaps his enemies had somehow stashed it away after his meeting with Bellarmine, only to produce it when it could do most harm to his case. Somewhere along the line, torture was mentioned. Not a nice thought, but then again, the Inquisition was not known for its niceness. Complexity, on the other hand, was something it relished, and the foregoing account of Galileo's interaction with the Church really only scratches the surface of what took place in the lead-up to the trial.

ON 22 JUNE 1633, THE INQUISITION ANNOUNCED ITS verdict. Guilty – and the specific crime: suspicion of heresy. This is an offence with three levels of seriousness: strong, vehement and slight. Galileo's accusers selected the intermediate level, but divided the heresy itself into two parts, each of which was addressed separately:

> You ... have rendered yourself according to this Holy Office vehemently suspected of heresy, namely
> (1) having held and believed a doctrine which is false and contrary to the divine and Holy Scripture: that the Sun is the centre of the world and does not move from east to west, and the Earth moves and is not the centre of the world, and
> (2) that [you] may hold and defend as probable an opinion after it has been declared and defined contrary to the Holy Scripture.

Galileo was duly sentenced:

> With a sincere heart and unfeigned faith, in front of us you [must] abjure, curse, and detest the above-mentioned errors and heresies, and every other error and heresy contrary to the Catholic and Apostolic Church, in the manner and form we will prescribe to you.
> Furthermore ... we order that the book *Dialogue* by Galileo Galilei be prohibited by public edict.
> We condemn you to formal imprisonment in this Holy Office at our pleasure.

It seems to be apocryphal that Galileo whispered the words *eppur si muove* (and yet it moves) after recanting his heresy. But it is certain that the sentence clipped Galileo's wings for the remaining nine years of his life. Imprisoned first at Siena in central Tuscany, he was eventually sent home to Florence, where he lived under house arrest. As his sight failed, he returned to the studies he had carried out before his work in astronomy, effectively inventing the new discipline of dynamics.

Although the Inquisition issued an edict forbidding the publication of his books, Galileo succeeded in bringing out one more, *Discourses and Mathematical Demonstrations Relating to Two New Sciences*, which was published in 1638 in the Protestant Netherlands. And, by the time of his death on 8 January 1642 at the age of 77, he had paved the way for Newton to develop his universal theory of gravitation. But it took another three and a half centuries – until November 1992 – for the Vatican to declare that Galileo had been right.

THE COMPLEX LEGAL TRAVAILS GALILEO EXPERIENCED were a direct consequence of his pioneering work with the newly invented telescope. Ideas that he knew to be true went against the accepted dogma of the age, and landed him uncomfortably in the spotlight of bigoted accusers. It was the first time a telescope had led to perceived breaches of the law – but far from the last. Throughout its 400-year history, the telescope has been the focus of some epic legal battles. Its evolution in the hands of gifted but sometimes unusual people has frequently thrust it into the centre of disputes, often as a result of technical developments – but sometimes quite unrelated.

Take, for example, the case of Richard Reeve, a London-based instrument-maker who produced the finest telescopes and microscopes available in Britain in the mid-1600s. By then, a number of improvements had been made to Galileo's telescope design. It was still a tube with a lens at either end, but the eyepiece lens – the one nearer the eye – had become a magnifying glass rather than the diminishing (concave) lens used by Galileo, an improvement that widened the field of view and rendered observation less like looking through a drinking straw. Yes, it turned the image upside down, but that was a minor detail for astronomy. And telescopes had become longer – very much longer, in fact. That was to counter a defect of 17th-century lenses that made them not only refract the incoming light to form an image, but also disperse it into rainbow spectrum colours, so that stars and planets seemed awash with coloured fringes.

Richard Reeve was a masterly optician, producing telescopes up to 18 metres long that provided high magnification with minimal false colour. They were used by the leading scientists of the time, including Robert Hooke, Robert Boyle and Christopher Wren – not to mention the rich and famous such as the diarist Samuel Pepys, who bought Reeve's instruments for both himself and his noble patrons. But Reeve apparently had a temper. In 1664, in a letter to Boyle, Robert Hooke wrote:

> Perhaps you may have heard of it: if not, in short, he [Reeve] has between chance and anger, killed his wife, who died of a wound she received by a knife flung out of his hand, on Saturday last. The jury found it manslaughter, and he had all his goods seized on; and it is thought it may go hard with him.

And at first it did, despite a subsequent note from Hooke that 'he now hopes that he will be able to get off, only it will cost him some money'. But what eventually transpired was a direct result of Reeve's skill as a telescope-maker. A few years earlier, he had made a 10.7-metre-long telescope for no less a personage than the king – the newly restored Charles II. And the king had been delighted with it. Could there be a connection between his delight and the royal pardon that was bestowed on Reeve some six months after his wife's death? It seems certain there was. The case was discharged, but it appears that it did, indeed, cost him a lot of money. The debt he incurred to his brother John, for example, is noted in John's will.

It was the eventual solution of the problem of spurious colour in telescope lenses – technically known as 'chromatic aberration' – that led to a much bigger legal spat in the annals of the telescope. The great Isaac Newton had declared the problem insoluble and turned his attention to the idea of using a dished mirror rather than a lens as the main image-forming component – the so-called 'objective'. That led to the first successful reflecting telescope, which he constructed in 1668. But a handful of individuals over ensuing decades wondered whether Newton might have been mistaken in abandoning the idea of colour-free lens telescopes. And the person who finally solved the problem in the early 1730s was not a scientist at all, but a barrister.

Chester Moor Hall worked at the Inner Temple, one of the four Inns of Court of the English judiciary. And he had an unusual hobby – the study of optics. After some experimentation, he devised a telescope objective that had two separate component lenses made of different types of glass. The idea worked, resulting in a lens that was 'achromatic', or free from spurious colour. Being a barrister, however, Moor Hall had no immediate use for his invention and, rather than patenting it, he simply passed it over to a couple of London telescope-makers he knew. But, unbelievably, after a few trials, they set it aside. Perhaps this newfangled telescope lens was just too hard to make, but the result was that for over two decades, the achromatic lens languished in obscurity.

It was a silk-weaver turned optician by the name of John Dollond who eventually rediscovered the idea, with a little help from an elderly jobbing optician who had made lenses for Chester Moor Hall back in the 1730s. In 1758, Dollond published an account of his experiments in the Royal Society's prestigious journal *Philosophical Transactions*. Then, urged on by his business-minded son, Peter, the elder Dollond successfully applied for a patent on the achromatic lens, allowing his company to flourish as the only legal manufacturer of optical instruments using it. Dollond's colour-free telescopes became the sensation of the age, with patrons including everyone from kings to Astronomers Royal, and unexpected luminaries such as President Thomas Jefferson and Mozart's father, Leopold – who was a noted amateur astronomer.

But other opticians in London were not impressed. They became aware that Chester Moor Hall had first invented the achromatic lens, and moved to challenge the Dollond patent. In a class action in 1764, thirty-five members of the Worshipful Company of Spectacle Makers petitioned the Privy Council to annul the patent, but were unsuccessful. Others simply ignored it, and produced achromatic telescopes of their own. But by then, John Dollond had died and Peter was the sole owner, taking a hard line on patent infringement. Several court cases ensued, including the case of *Dollond vs. James Champneys of Cornhill, London*, in which the Chief Justice of the Court of Common Pleas, Lord Camden, noted with regard to Moor Hall's invention that ' … it is not the person who locks his invention in his scrutoire who ought to profit by a patent for such invention, but he who brings it forth for the benefit of the public'. Champneys and many others wound up paying crippling damages and royalties that quickly sent them bankrupt, while the Dollond company went from strength to strength. Until 2015, the name could still be seen in the British high street optical chain of Dollond and Aitchison, a company that lives on today under the Boots Opticians brand name.

A CENTURY AFTER CHESTER MOOR HALL'S EXPERIMENTS, the achromatic lens again became the centre of a legal dispute.

This time, however, it involved two of the highest profile figures in British astronomy. In 1829, Sir James South and the Reverend Richard Sheepshanks became founding president and secretary of what was soon renamed the Royal Astronomical Society. The problem with these two strong-minded individuals was they had differing views on almost everything, and frankly despised each other. Their animosity boiled over in legal proceedings after South had purchased an exquisite 30-centimetre-diameter achro-matic lens from a noted French optician, with the aim of pursuing his studies of double stars. This was the largest telescope lens in Britain at the time, and South hired a well-respected instrument-maker, Edward Troughton, to build the telescope to house it. Things did not go well and, in 1832, after two and a half years of work, South wrote to Troughton, accusing him of delivering 'a useless pile'. Well out of pocket, Troughton took legal action against South, hiring a certain lawyer who was also a mathemati-cal genius and an ordained minister in the Church of England: one Richard Sheepshanks.

Six years of legal wrangling followed, but in 1838, matters were resolved in Troughton's favour. Sadly, he had died three years earlier, but his company was awarded £1470 in costs against Sir James South. This tipped South over the edge, and no doubt Sheepshanks' involvement in the case enraged him far more than the financial loss. In 1839, he took to the unfinished telescope with an axe, putting up posters all over London advertising the sale of its dismal remnants, and decrying Troughton, Sheepshanks and their accomplices (who included no less a personage than the Astrono-mer Royal). And he repeated the exercise with the remaining bits and pieces in 1843.

The feud between South and Sheepshanks raged on for another decade until Richard Sheepshanks passed away in 1855. But that wasn't quite the end of it, as South took a verbal swipe at the Royal

Astronomical Society's glowing obituary for his old enemy. The saddest part of the story is that the magnificent French lens never realised its full potential. By the time it was finally mounted in a telescope, at the University of Dublin's Dunsink Observatory in 1863, it was a small instrument by the standards of the day. It's still used for amateur astronomy and teaching.

WHILE WE MAY SMILE AT THE LIKES OF SOUTH AND Sheepshanks, there is really nothing funny about the legal disputes that sometimes surround telescopes today. As we noted in chapter 2, modern optical telescopes are major international collaborations, located on high mountain-tops where atmospheric conditions are superbly matched to astronomers' requirements. Since the early 20th century, reflecting telescopes equipped with large dished mirrors have overtaken lens telescopes as the instruments of choice, simply because mirrors can be made bigger. And in astronomy, size is everything, to maximise the light gathered from faint objects in deep space. Today's biggest telescopes have mirrors 8 to 10 metres in diameter, sometimes made of a single piece of high-tech glass, but often composed of interlocking hexagonal segments held in perfect alignment by computer-controlled fingers. They are known generically as 'very large telescopes' and indeed, among the most productive of them is a quartet of 8.2-metre instruments operated in northern Chile by the European Southern Observatory (ESO), known collectively as the VLT.

But we are now on the brink of a new generation of ELTs, or extremely large telescopes, with mirrors 20 metres or more in diameter. And they have brought to a head some serious legal issues surrounding their construction. In fact, what is happening today was foreshadowed during the 1990s, when the University of Arizona embarked on constructing an international optical observatory on Mount Graham in the south of the state. This tree-covered 3200-metre peak is home to the endangered Mount Graham Red Squirrel and, to the discomfort of astronomers (who are by nature environmentalists), a highly publicised legal challenge took place.

The dispute of astronomers vs. conservationists quickly escalated to incorporate the question of the mountain-top's traditional ownership. Like other peaks in the Pinaleño Mountains, Mount Graham is a sacred site for the San Carlos Apache people. After years of protests and legal wrangling, a compromise was eventually reached with the Apache Tribal Council, and construction of the observatory was approved by an Act of the US Congress, with the proviso that an independent census of the squirrel population should be carried out. The biggest telescope on the mountain, the 2 × 8.4-metre Large Binocular Telescope, was eventually built between 1998 and 2004, coinciding initially with an unexpected peak in squirrel numbers, but now seeing numbers similar to those before the site was developed for astronomy.

Aside from environmental issues, the clear lesson from this episode is that the mountain peaks favoured by astronomers for their giant telescopes are often deeply significant for the indigenous people of the area. And of the three ELTs currently under development, one is now embroiled in a conflict whose outcome is difficult to foresee. This is the TMT, or Thirty Meter Telescope project, whose multinational proponents expect the telescope to be built on land managed by the University of Hawaii on the 4200-metre summit of Mauna Kea on Hawaii's Big Island. (The other two ELTs are in the southern hemisphere, by the way, and already under construction on sites in northern Chile. They are ESO's 39-metre European ELT and the 23-metre Giant Magellan Telescope, both of which have the approval of indigenous authorities.)

In fact, development on Mauna Kea has been controversial since the first astronomical facilities were built there in the late 1960s. The summit area is sacred in native Hawaiian religion, and is visible from virtually the whole island. Today, there are 12 separate facilities on the mountain, reflecting its status as the northern hemisphere's best site for optical astronomy. However, the thirteenth – the giant TMT – will be enclosed in the most visible structure by far, and has triggered unprecedented protest. In October 2018, after seven years of controversy, the Supreme Court of Hawaii approved construction of the telescope, and that is legally where things stand today. However, a public forum in March 2019 drew bitter criticism from the local community, with the University of Hawaii being accused of '50 years of mismanagement', and the Native Hawaiian Legal Corporation wondering what has become of the spirit of aloha in the dispute.

In a conflict that seems light years from Galileo's trial, there are no winners over this sensitive issue. Astronomers clearly want to respect the traditional culture of Hawaii, but are jealous of the superb conditions that nature has dealt them on Mauna Kea. The best that can be hoped for is a compromise that will probably involve decommissioning some existing facilities as a gesture of goodwill, and perhaps a review of the site's management structure. As someone who has cherished the pristine skies of Mauna Kea since my first visits there 40 years ago, I sympathise with all parties in the dispute. I'll be watching developments with interest.

CHAPTER 9
SPACE BUGS: RULES FOR PLANETARY PROTECTION

This story starts with an auspicious day in the annals of spaceflight – Wednesday, 19 November 1969, during the unforgettable era of NASA's lunar landings. A few minutes before 7 am GMT on that day, *Apollo 12* became the second human-occupied spacecraft to touch down safely on the surface of the Moon, after the historic landing of *Apollo 11* the previous July. Of course, I'm quoting Greenwich Mean Time here, rather than Houston time, because that was the time zone in which I was glued to a blurry TV screen in wintry Scotland, watching every last detail of the astronauts' extra-vehicular activity – or moon-walk, to you and me.

Despite my attentiveness, I managed to miss one of the most memorable events of the mission. The pinpoint landing had brought *Apollo 12*'s astronauts Pete Conrad and Alan Bean to within walking distance of a robotic spacecraft that had been sitting on the Moon's surface for two and a half years. *Surveyor 3* was part of NASA's intensive preparation for the Apollo flights back in April 1967, and mission scientists wanted to investigate the effect of long-term exposure to the harsh lunar environment. What would the near-complete vacuum, monthly temperature range of

–150 °C to +120 °C, and relentless bombardment by subatomic particles do to the fragile components of a spacecraft?

Conrad and Bean duly removed various bits and pieces from *Surveyor*, including its TV camera, and packed them up for the return to Earth. Actually, that sounds a bit more meticulous than it really was, because the camera was stuffed into a nylon duffle bag rather than one of the special airtight boxes normally reserved for lunar samples. Nevertheless, it made a safe splashdown in the South Pacific Ocean on 24 November 1969 in *Apollo 12*'s Command Module, along with Conrad, Bean and pilot Dick Gordon.

For the camera, though, that was just the start of the story. Scientists examining it after its return were surprised to find spores of a common bacterium, *Streptococcus mitis*, residing in its insulating foam. These little critters are found in the mouths and throats of humans. When the spores were cultured, they proved to be perfectly viable, leading to the remarkable conclusion that microbes deposited on the camera before lift-off – perhaps by someone sneezing on it – had survived on the Moon for more than two years. This conclusion, published in the academic literature in 1971, was regarded by Pete Conrad as 'the most significant thing that we ever found on the whole ... Moon'.

More recently, however, the claim has been disputed, since there is evidence that the contamination may have occurred after the camera left the Moon. A 'breach of sterile procedure' in the lab has been cited, as well as the possibility that the camera was contaminated while it was in its duffle bag in the *Apollo 12* Command Module – in close proximity to the three returning astronauts. It does seem hard to imagine that none of them sneezed during the three-day trip back to Earth. On the other hand, there are aspects of the culturing results that suggest the bacteria were present on the Moon. For example, they took some time to spring into life, and they were only found within the insulating foam rather than on its surface – both of which would be unlikely to happen with later contamination. The bottom line is that we will probably never know the truth, but the episode did highlight the possibility that microbes could survive the rigours of space.

SINCE THE *APOLLO 12* EPISODE, THERE HAVE BEEN OTHER examples of space-hardy microbes coming back to Earth with a decided zest for life. Perhaps the most memorable is the 553 days endured on the outside of the International Space Station by a batch of terribly British microbes from Beer. (Not the beer you drink, but a fishing village in Devon that got its unusual name from the Old English word for a grove of trees – *bearu*.) A few lumps of rock from the cliffs of Beer – complete with their microbial inhabitants – were mounted onto the outside of the space station in 2008. Eighteen months later, they were returned to Earth and examined. Why?

Scientists from the Open University in the United Kingdom, who were responsible for the experiment, wanted to test the effects of the space environment on a completely random sample of microbes, rather than preselecting the ones that we already know are pretty hardy. Those are usually known as extremophiles, because of their lust for extremes, and include organisms that can survive below freezing point, as well as ones that don't mind being in boiling water. The Beer microbes weren't extremophiles; they were just ordinary workaday microbes, constituting several different communities of micro-organisms. What the scientists wanted to know was which of these are sufficiently hardy in space to be useful to future space explorers in naturally recycling waste products in life-support systems. And now they know – because a sizeable fraction of the population survived to tell the tale back on Earth. It could be a very useful discovery, and certainly bolsters our understanding of the way single-celled organisms react to the harshness of space.

But – amazingly – there is at least one *animal* species that can survive the vacuum and radiation environment of space. Meet the tardigrade, an eight-legged animal that is also known as a water-bear – although with a maximum length of a millimetre, this is not a bear you'll ever have to flee screaming from. Some of the thousand or so known species of tardigrade are able to survive incredible extremes of temperature, pressure and radiation: they are simply the toughest animals on the planet. They survive by shutting down their metabolism and curling up into a dehydrated ball. On Earth, specimens have endured for decades in this state, before being successfully rehydrated in water. The space tardigrades, lofted on 14 September 2007 in a European experiment called Biopan-6, survived the rigours of open space for ten days. Or, at least, some of them did. Survival rates were not particularly high, but several of the returning spacefarers did go on to produce perfectly normal offspring. As the prestigious scientific journal *Nature* cheerfully reported, for these creatures, space suits are optional.

The evident space-hardiness of common earthly microbes raises all kinds of interesting possibilities. For example, did microbial life come to Earth from elsewhere in the Solar System? Panspermia ('seeds everywhere') is an old idea reinstated several decades ago by the iconoclastic British astronomer Fred Hoyle, in collaboration with Chandra Wickramasinghe, now of the University of Buckingham. These eminent scientists developed a theory of cometary panspermia in which rudimentary life-forms are common throughout the Universe, and travel through space from one planet to another. The idea remains wildly controversial, but we do now know that carbon-containing molecules important for life processes were present in the cloud of gas and dust from which the Solar System formed 4.6 billion years ago. They are preserved today in comets, and suggest that perhaps the building blocks of life, at least, came from space. As unlikely as it might be, the cometary panspermia idea cannot yet be ruled out.

THE TWO DIFFERENT SITUATIONS OF EARTHLY MICROBES being transported to another celestial body (as might have occurred with *Surveyor 3*) and the transfer of extraterrestrial organisms (if they exist) back to Earth have special significance in the field of Solar System exploration. They are referred to, respectively, as 'forward contamination' and 'back contamination' – rather unimaginative terms that nevertheless highlight the need to take care whenever space missions are being planned.

Why are they so important? Forward contamination could, conceivably, introduce earthly microbes into an extraterrestrial environment where they could flourish to the detriment of any hypothetical indigenous organisms. Maybe the results of that contamination wouldn't be apparent for a few million years, but clearly it would still be a Bad Thing. And the consequences of back contamination are just as unpredictable. Okay, maybe a few malicious Martian microbes accidentally brought home on a future sample-return mission wouldn't wipe out all life on Earth, but we simply don't know.

The bottom line is that the space world takes these possibilities extremely seriously – and has for years. The issue of potential planetary (or lunar) contamination was first raised as long ago as 1956 at an Astronautical Federation Congress. Not only did this predate the pioneering orbital flight of *Sputnik 1* in 1957, but it occurred long before there was any real evidence that microbes could survive in space. Then, in 1967, the United Nations Outer Space Treaty was ratified, providing the foundations of space law.

With great foresight, it incorporated a set of so-called planetary protection rules that are still in use today.

Article IX of the Treaty provides the legal basis. It states that: 'Parties to the Treaty shall pursue studies of outer space, including the Moon and other celestial bodies, and conduct exploration of them so as to avoid their harmful contamination and also adverse changes in the environment of the Earth resulting from the introduction of extraterrestrial matter and, where necessary, shall adopt appropriate measures for this purpose.' Bold words, but increasingly significant ones in an era in which we know contamination is a real possibility.

The planetary protection rules are now managed by the multinational Committee on Space Research (COSPAR), a large group of scientists, which meets every two years. COSPAR defines different categories of missions, ranging from Category I (any mission to locations not of direct interest for chemical evolution or the origin of life, such as the Sun, or Mercury) to Category V, which is concerned with sample-return missions that could bring extraterrestrial biological materials to Earth.

The other categories defined under the planetary protection rules sit fairly logically between these limits. Category III is for fly-by and orbiter missions to 'locations of significant interest for the chemical evolution and/or the origin of life', with a significant chance that contamination could compromise future investigations. Such locations include Mars, of course, and some places in the outer Solar System such as Jupiter's moon Europa, and Saturn's moon Enceladus. Finally, and most importantly, Category IV is for spacecraft that will actually land in such locations.

BECAUSE MARS IS OF SUCH GREAT INTEREST IN THE SEARCH for life beyond Earth, there are special rules that apply to the red planet. In fact, COSPAR defines Mars Special Regions as those within which terrestrial organisms could readily propagate, or those that are thought to be more likely to host Martian life-forms.

In particular, any region of Mars in which liquid water could occasionally occur (and there are a few, despite the planet's sub-zero average temperature) are classified as Special Regions. They include a 20-kilometre-wide lake recently discovered under the ice of the southern polar cap. These places are subject to the most stringent planetary protection rules, the so-called Category IVc. This states that sterilisation must be achieved to a maximum of 30 spores per spacecraft. To a non-biologist like me, that sounds like an ultra-low level of contamination, and, indeed, is known as the '*Viking* post-sterilisation biological burden', because NASA's two *Viking* landers of 1976 were sterilised to this level. It was achieved by baking each entire spacecraft at a temperature of nearly 112 °C, and then enclosing it in a pressurised cocoon known as a 'bioshield' to prevent any biological contamination until the spacecraft had left Earth's atmosphere. The bottom line, though, is that sterilisation incurred a cost of approximately US$100 million out of a total mission cost of US$1 billion.

This level of expense on lander missions to Mars has led some astrobiologists to argue that the Category IVc rules should be relaxed. One prominent scientist (Ryan Anderson, who works on NASA's *Curiosity* rover) has remarked that it's paradoxical that the most habitable parts of Mars are the toughest places to send new spacecraft to. In any case, it's possible that Martian organisms might already have found their way to Earth on meteorites that are known to have travelled between the two planets – a variation of Hoyle and Wickramasinghe's panspermia hypothesis. Some scientists have even suggested that terrestrial life actually originated on Mars perhaps four billion years ago. Others who support relaxing the rules claim that there will be no problem in determining the origin of any Martian microbes we may find, once a robotic DNA sequencer has been sent to the planet in some future mission.

Most scientists, however, support the status quo, arguing that any contamination of Mars makes the task of finding putative Martian life more difficult. Significantly, they also note that if such life is contaminated by terrestrial organisms, it's a one-way process – there's no possible return to the pristine state.

Mars, of course, is also a target for human exploration, which will bring its own contamination issues. NASA currently plans to have astronauts walking on the surface of the red planet in the mid-2030s, a date that is more likely to be delayed than brought forward, due to the enormous technical challenges of such a mission.

I recently had the opportunity of asking two NASA luminaries – an astronaut and an astrobiologist – how the Category IV rules will be applied when crewed missions go to Mars, since complete sterilisation is clearly impossible. Despite the fact that these individuals worked in completely different areas of NASA, and were on opposite sides of the globe when I spoke with them, their answers were remarkably similar. There was an underlying assumption that robotic missions between now and the crewed landing will fail to find any signs of life. If that turns out to be the case, then perhaps the Category IV rules could be loosened a little, to allow microbe-riddled humans to visit. But the bigger surprise came when I asked what might happen if living organisms are found there. Both lowered their voices and adopted a similarly conspiratorial air to tell me that 'Well, the planetary protection rules will probably be quietly dumped.' It remains to be seen just how this will play out down the track.

MARS IS NOT THE ONLY PLACE WHERE STRINGENT ANTI-contamination procedures are applied. Other hot spots in the search for life in the Solar System are some of the moons of the giant planets. We know that Jupiter's moons Europa, Callisto and Ganymede have a rocky core overlain by a global ocean of liquid water, which is itself overlain by a thick crust of ice. Saturn's moons Titan, Enceladus and Dione are thought to have a similar structure.

Intriguingly, both Europa and Enceladus have spectacular geysers of ice crystals erupting from their south polar regions, offering free samples of the subsurface ocean to any properly equipped spacecraft that can fly through them. While the NASA/ESA/ASI *Cassini* spacecraft was exploring Enceladus, we found chemical evidence of hydrothermal vents on the floor of the Saturnian moon's sub-ice ocean. This is highly suggestive, since similar active vents on the infant Earth's ocean floor are thought to be one of the places where life originated on our own planet.

Arguably, Saturn's moon Titan has even more to offer. As well as a liquid water ocean underlying its hard-as-rock ice surface, it has frigid seas and lakes of liquid hydrocarbons on top. They are effectively seas of liquid natural gas, and they are in equilibrium with Titan's thick atmosphere, replenished from time to time by heavy showers of oily rain. Moreover, the seas and lakes could harbour life-forms based not on water (as all life on Earth is), but on the constituent chemicals of natural gas – methane and ethane. We'll visit this extraordinary world again in chapter 13.

Any living organisms in the sub-ice oceans of Europa and Enceladus, or in the hydrocarbon seas of Titan, are likely to have originated quite independently of life on Earth. The distances involved are huge, and, in Titan's case, any hydrocarbon-based life would be totally different from earthly life. Thus there is much at stake in risking contamination. For this reason, NASA's *Galileo* probe, which studied Jupiter in the early 2000s, was made to burn up in the atmosphere of the giant planet in 2003 to avoid any possible contamination of its moons. Likewise, the highly successful *Cassini* probe was intentionally destroyed by having it enter Saturn's atmosphere on 15 September 2017.

The future exploration of the Solar System's ice moons without contaminating their surfaces presents particular problems to planetary scientists. Perhaps that's why most currently proposed missions stick to exploration from orbit, rather than risking a landing. Hence *JUICE* (Jupiter Icy Moons Explorer) is ESA's spacecraft to Europa, Callisto and Ganymede, scheduled to begin an eight-year journey in 2022. Its sojourn there will allow mission scientists to take a close-up look at those worlds with the possibility of sub-ice life-forms firmly in mind.

NASA has plans, too. *ELF* (Enceladus Life Finder) will, if eventually approved by NASA, make multiple flights through the ice fountains of the Saturnian moon looking for suggestive molecular signatures. Whereas *ELF*'s predecessor, *Cassini*, could detect inorganic molecules like hydrogen and silicates, *ELF* will be looking for biological precursors such as nucleic acids, amino acids and lipids. And some life-related molecules can be detected through the submillimetre radio waves they emit, which has led to another NASA proposal called *SELFI* – the Submillimetre Enceladus Life Fundamentals Instrument.

As with all similar missions, *JUICE*, *ELF* and *SELFI* would require Category III sterilisation, prior to their long journeys to the gas giants. And their travels would almost certainly end with suicidal plunges into the respective parent planets, to avoid contaminating the moons.

CHAPTER 10
CLIMATE CHANGE: WHAT HAPPENED TO MARS?

O f all the planets of the Solar System, the best studied is also the one most fantasised about. Mars has captured the popular imagination since astronomers began speculating that it might be a world like Earth, soon after the invention of the telescope. By the end of the 19th century, those fantasies had reached fever pitch, with suggestions by Italian astronomer Giovanni Schiaparelli and American astronomer Percival Lowell that an advanced civilisation must have excavated a planet-wide network of irrigation channels ('canals') in the face of global climate change. Dark markings visible using ground-based telescopes were obviously areas of vegetation fringed by encroaching Martian deserts, and fed by water artificially channelled from the planet's icy polar caps. It was only a matter of time before we would be communicating with the Martians themselves.

All very neat and tidy, until the *Mariner 4* fly-by of July 1965 (and subsequent *Mariner* and *Viking* missions) revealed that almost everything we thought we knew about Mars was wrong. With its cratered surface, a dry and windy atmosphere only 1 per cent as dense as Earth's that stirs up frequent dust storms, and its frigid surface temperature (–65 °C on average), the new Mars was a decidedly inhospitable place. And 50-odd years of subsequent research, carried out with a flotilla of Mars orbiters and landers, has done nothing to change this view.

All the Earth's attributes that make it a benign environment for life to evolve are absent on Mars. There's no global magnetic field to shield the planet from the solar wind. There's no green-house blanket of air to moderate the surface temperature – and nothing to regulate its carbon content, as there is on Earth. And the absence of a massive moon renders the planet's axial tilt unstable. Yet Mars shows tantalising signs of having been very different in the past, and that suspicion drives today's research efforts.

THE FACT THAT MARS HARBOURS THE LARGEST VOLCANIC plateau in the Solar System – the Tharsis Rise – shows that it was once a geologically active planet. Five huge volcanoes dominate the Tharsis region, one of which (Olympus Mons) is the biggest in the Solar System. These are shield volcanoes, similar in structure to those that make up the Hawaiian Islands. Their shallow slopes come from low-viscosity magma. The summit caldera of Olympus Mons stands a whopping 27 kilometres above the surrounding landscape. That extraordinary elevation is thought to be due to the absence of plate tectonics on Mars. Like an orange, the planet has an unbroken skin – a single crustal plate that may have remained stationary over a hotspot in the underlying mantle for a very long time, allowing Olympus Mons to grow to its gargantuan size.

While the growth of the Tharsis region continued throughout the most recent geological era of Mars (known as the Amazonian period, which started about 2.9 billion years ago), it is the earlier history of the planet that tantalises planetary scientists. In the oldest, or Noachian era, dating from the planet's formation 4.6 billion years ago and lasting some 900 million years, the planet was undoubtedly warm and wet. Ancient clays and sedimentary rock formations dating from the Noachian are widespread on Mars.

The hard evidence that liquid water was abundant comes both from orbiting spacecraft like NASA's *Mars Reconnaissance Orbiter* and rovers such as *Spirit, Opportunity* (both now defunct) and *Curiosity*. The orbiters look at the big picture, with cameras, radar and analytical instruments that provide coverage of the whole planet. The rovers, on the other hand, get up close and personal with the Martian surface, acting as robotic geological laboratories equipped to investigate every aspect of its rocks and soil. *Curiosity* even has a 'laser zapper' called ChemCam, which can sense the chemical composition of rocks up to 7 metres away by analysing the light emitted when the laser vaporises small areas of their surfaces.

Geographical features associated with water erosion are supported by evidence from soil and rock analysis, which reveal minerals that only form in the presence of liquid water. Moreover, the presence of gravels containing smooth pebbles demonstrates that rivers and streams flowed for significant periods of time at *Curiosity*'s landing site in a geological wonderland known as Gale Crater (named after the Australian amateur astronomer Walter Frederick Gale, whose discoveries during the late 19th and early 20th centuries included comets, double stars and Martian features that, like Schiaparelli and Lowell, he believed to be canals).

Further results from *Curiosity*'s analysis of an ancient lake bed show that it was laid down in fresh water, rather than the acidic brine that gave rise to the clays *Opportunity* had analysed earlier in a different region of the planet. That brine was much saltier than the Earth's oceans (though not as saline as the Dead Sea), and the presence of an iron sulphate mineral called jarosite suggested acidity, as jarosite only forms in such environments. The acidity is also cited as a reason why the ancient Martian seabeds are not rich in carbonates, as terrestrial ocean beds are.

Most tantalising of all is the suggestion that the whole of the low-lying northern hemisphere of Mars was once covered by water. In high-resolution images of the planet taken from orbit, we see features normally associated with water erosion here on Earth: river valleys, oxbows, canyons, outwash flows, and evidence of beaches and sea cliffs along what is now taken to have been an extensive coastline. Laser altimetry from orbiting spacecraft (especially NASA's *Mars Global Surveyor*, operational from 1997 to 2006) has shown the northern hemisphere of Mars to be flatter, lower-lying, and less cratered than its southern counterpart, leading to the idea that it once harboured an ocean. An earlier objection that the supposed shoreline varied in height around the ocean rim (and therefore couldn't be a shoreline) has been refuted with the suggestion that large-scale shifts in the inclination of Mars' rotation axis occurred, due, perhaps, to eruptions of Olympus Mons.

The question of whether the 5-kilometre height dichotomy between the northern and southern hemispheres is the result of an oceanic basin – or of some other cause such as convection in the planet's mantle or a major asteroid impact – is still controversial. While it is generally accepted that a stable body of water did cover most of the northern hemisphere, the jury is still out on how deep it was, and how long-lasting. Were there tens of millions of years of constant cover, or wet episodes interlaced with long periods when the seabed was dry? Either way, the demonstrated existence of liquid water in the ancient Noachian era is an exciting find for astrobiologists, whose studies of the prospects of life having arisen elsewhere in the Universe invariably begin with a watery environment – because all life on Earth uses water as its working fluid.

It is believed that wet conditions on Mars lasted well into the Hesperian era, which occurred between 3.7 and 2.9 billion years ago. This is the period during which we know life was beginning on Earth, the oldest undisputed fossilised terrestrial bacteria dating from three billion years ago, with more controversial evidence of micro-organisms existing half a billion years earlier.

Thus it is that the search for life beyond Earth is entering a critical phase. Soon after it arrived on Mars in 2012, *Curiosity* achieved its mission's primary goal — to discover whether Mars was ever habitable. Having established that, it now remains for us to find whether that ancient habitability actually spawned living organisms. And, if it did, to discover what happened to them. We'll pick up that story again in the next chapter, but there is one further intriguing prospect that relates to the panspermia theory outlined in chapter 9. Is it possible that Martian micro-organisms could have been the source of life on Earth, having made their interplanetary journeys on the meteorites that are known to have travelled between the two planets? Did life on Earth share a common origin with its putative Martian counterpart? This is just one of the many possibilities astrobiologists are investigating today.

SO, IF MARS DID HAVE A WET PAST, WHAT HAPPENED TO change it, and is there a lesson in climate change for we dwellers on planet Earth? The geological evidence points to the Martian sea or ocean having disappeared between two and four billion years ago — that is, somewhere in the first half of the planet's 4.6 billion-year lifetime. And the trigger seems to have been its small size — about half the diameter of Earth. With a proportionately smaller iron core, Mars had only a limited reservoir of internal heat left over from its fiery birth and, while it is thought that the core remains at least partly liquid, it is no longer hot enough to sustain either an internal dynamo or plate tectonics. It seems likely that these processes are long gone, having shut down during the ancient Noachian period.

The motion of rock plates in the Earth's crust plays an important role in stabilising the atmosphere, because it circulates carbon between the atmosphere and the mantle beneath. But as Mars' molten core cooled more rapidly than Earth's, plate tectonics shut down early in its history, removing the 'thermostat' that allowed carbon dioxide to keep the planet warm. Thus the planet lost most of its greenhouse blanket, gradually cooling to become the frigid world we see today. The cooling core on Mars is also the reason the planet has no appreciable magnetic field, resulting in unmitigated exposure to the solar wind. That would have enhanced the process of atmospheric water vapour being dissociated into hydrogen and oxygen – and lost to space.

That's not to say that Mars is now devoid of water, however. Much of it is still there, locked up as ice in the polar caps, or beneath the surface soil as permafrost at lower latitudes. Ground-penetrating radar aboard orbiting spacecraft has revealed glaciers overlaid by a thin layer of soil, even at temperate latitudes. And during its six-month mission in 2008, NASA's *Phoenix* lander discovered a permafrost of ice only millimetres beneath the surface soil in the Martian arctic. It also demonstrated that a limited water cycle exists between atmosphere and ground, with occasional observations of snowfall. By contrast, the overall quantity of ice on Mars is far from limited. Data from ESA's *Mars Express* orbiter has revealed that if just the southern polar cap were melted, it would produce enough water to flood the entire planet to an average depth of 11 metres.

The fact that Mars was once a habitable planet but isn't now highlights how delicate the balance of the Earth's atmosphere is. So the lesson for we Earth-dwellers is: don't tinker with it. Especially if it involves plate tectonics.

CHAPTER 11

NOT OUR PLANET B?
COLONISING MARS

When you think about the possibilities of life in the Solar System, it's hard to avoid the conclusion that we humans are top dog. Of course, the sub-ice oceans of distant worlds such as Jupiter's moon Europa or Saturn's Enceladus could harbour super-intelligent beings, just waiting for their moment to enslave the inhabitants of Earth. But to be honest, that seems pretty unlikely.

So, here we are, a technological species with the trappings of civilisation, and we think we're pretty clever. I must admit that every time I sit in an aircraft flying above three-quarters of the Earth's atmosphere at close to the speed of sound, I marvel at what we can achieve with technology. And that's just the commonplace: when I think of robotic spacecraft festooned with intelligent sensors exploring the Solar System, I'm in awe of the way scientific curiosity can draw such extraordinary feats of engineering out of the minds of humans.

We do have our failings, of course. There's a dangerous enthusiasm for our vestigial tribalism, which means we have to maintain expensive armed forces. We also have a propensity for trashing the environment that sustains us – and that's something we're really good at. And because we are so good at it, some of us have already written off our planet as a lost cause. Perhaps there's little wonder, then, that those folk are looking to the heavens for salvation. And where is their rapacious gaze focused? Firmly on the fourth rock from the Sun.

MARS IS A WORLD WITH EXTRAORDINARY PARALLELS TO our own. It has a rocky surface, with a generally transparent atmosphere. While it's only half the diameter of Earth, its day-length (24 hours 47 minutes), axial tilt (25.2 degrees) and land area (145 million square kilometres) are remarkably similar. (The latter arises because Mars has no oceans.) Okay, its average daytime temperature is around −40 °C, and its atmospheric pressure is only about 1 per cent of ours, but the great thing about Mars is that it's empty. Or at least, it seems to be. And that's why it's widely touted as Earth's lifeboat – the place that humans could inhabit as their insurance against a civilisation-ending catastrophe on Earth.

What sort of catastrophe are we talking about? It could be our own fault. Climate change, overpopulation, global war – to name just the obvious ones – but perhaps our more likely demise is at the hands of a runaway virus, a supervolcano, a rogue asteroid or a nearby exploding star. I remain an optimist when it comes to our survivability in self-inflicted catastrophes, given how resilient and inventive humans are. And most natural threats have technological solutions, too, if we can act quickly enough. Except, of course, the last one on my list, in which it wouldn't make any difference whereabouts in the Solar System you were. You'd still be fried.

In this context, it's of interest to ask whether humankind has ever come close to such catastrophic extinction in the past. Conservation biologists look at what are known as genetic (or population) bottlenecks, in which the size of a population suddenly falls, due to some external agent. That might include environmental changes such as famines caused by drought, floods or natural catastrophes

A spectacular display of antisolar crepuscular rays, seen near Canberra at sunrise. The rays appear to converge towards a point directly opposite the Sun, which is shining through clouds behind the photographer.

Marnie Ogg

Topped by the purple-pink 'Belt of Venus', the Earth's shadow rises over the eastern horizon soon after sunset at Narrabeen Lagoon in northern Sydney. The shadow is called the 'twilight wedge' because of its three-dimensional shape in the atmosphere.

James Watson

Celebrity physicist Brian Cox is dwarfed by the 3.9-metre-diameter
Anglo-Australian Telescope during *Stargazing Live* in 2017. Sited at Siding
Spring Observatory, the telescope is the largest of its kind in Australia.

Ángel López-Sánchez

A 14-centimetre-long nickel-iron meteorite, photographed on the sands of Gale Crater on Mars by NASA's *Curiosity* rover. Three tiny dots reveal where *Curiosity*'s laser has probed the meteorite's chemical composition.

NASA/JPL-Caltech/MSSS

Above Sentinel at dawn. Pan-STARRS1 searches for potentially hazardous asteroids from its mountain-top vantage point on the Hawaiian island of Maui. In the distance is the 4200-metre summit of Mauna Kea on the Big Island, 125 kilometres to the south-east.

Rob Ratkowski/University of Hawaii/STScI-H-p1912a-f

Top right Antennas of the Australian Square Kilometre Array Pathfinder at the Murchison Radio-Astronomy Observatory, Western Australia. Wajarri Yamatji elder Ernie Dingo likened them to 'beautiful giant white wildflowers growing up out of the earth'.

Author

Below right Not a lift-off but a set-down. In a beautifully choreographed manoeuvre, two of the three boosters of SpaceX's Falcon Heavy land at Cape Canaveral for reuse after the rocket's first commercial launch in April 2019. The third booster landed safely at sea.

SpaceX

Above No space suit needed. The tardigrade, or water bear, is a common invertebrate that can survive the vacuum of space. This colour-enhanced scanning electron micrograph of a tardigrade in its mossy environment has a magnification of 400x.

Science Photo Library

Top left The Deep Space Climate Observatory (DSCOVR) monitors space weather from its location between Earth and the Sun. Here it has also captured the Moon's far side, revealing how dark our satellite is compared with Earth.

DSCOVR NASA/NOAA

Below left A sunset view of the European Southern Observatory's Very Large Telescope complex at Cerro Paranal in northern Chile. Each of the large enclosures contains an 8.2-metre telescope that can be used independently or with its companions, along with the smaller domes in the foreground.

G Hüdepohl ESO

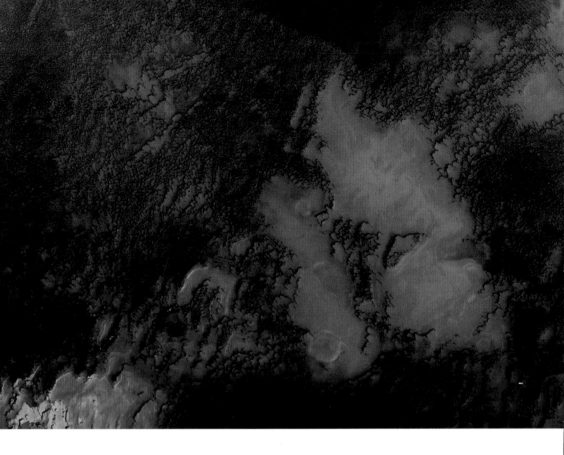

Above The front of a Martian dust cloud photographed in the planet's northern arctic in April 2018 by the European Space Agency's *Mars Express* orbiter. It heralded the onset of a particularly intense dust storm season, which soon brought a Mars-wide storm.

Mars Express, ESA/DLR/FU Berlin

Right A grubby-looking *Curiosity* used its Mars Hand Lens Imager to mosaic this self-portrait during a lull in the dust storm of June 2018.

NASA, JPL-Caltech, MSSS

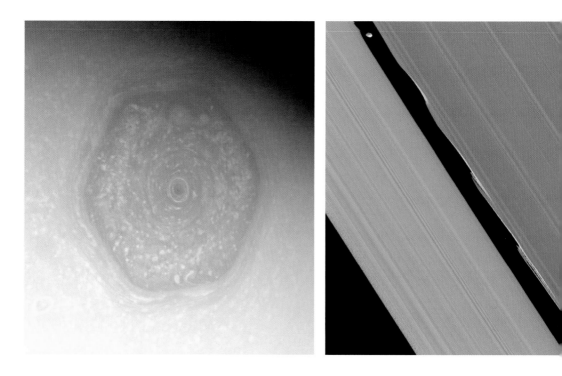

Top left NASA imaging expert Val Klavans' true-colour portrait of Saturn's north polar hexagon uses monochrome imagery taken by the *Cassini* spacecraft in June 2013. The extraordinary shape results from waves in the planet's polar jet stream.

NASA/JPL-Caltech/Space Science Institute/Val Klavans

Top middle The 42-kilometre Keeler Gap is close to the outer edge of Saturn's main A-ring, and is caused by tiny Daphnis, an 8-kilometre-wide moon. Daphnis is at the upper left in this remarkable *Cassini* image, which also shows waves in the ring boundary induced by the moon's gravity.

NASA/JPL-Caltech/SSI/Kevin M. Gill

Top right Back-illuminated by the Sun in this 2009 *Cassini* image, Saturn's moon Enceladus reveals the fountains of ice crystals that spray from its frozen surface. Originating in a global ocean, the fountains may contain signs of life. The rings and Saturn's small moon Pandora are also visible.

NASA/JPL-Caltech/Space Science Institute

Bottom right Saturn's giant moon Titan lurks behind the ice moon Rhea in another striking *Cassini* image. The difference is unmistakable, Titan's opaque atmosphere contrasting with Rhea's heavily cratered surface.

NASA/JPL-Caltech/Space Science Institute

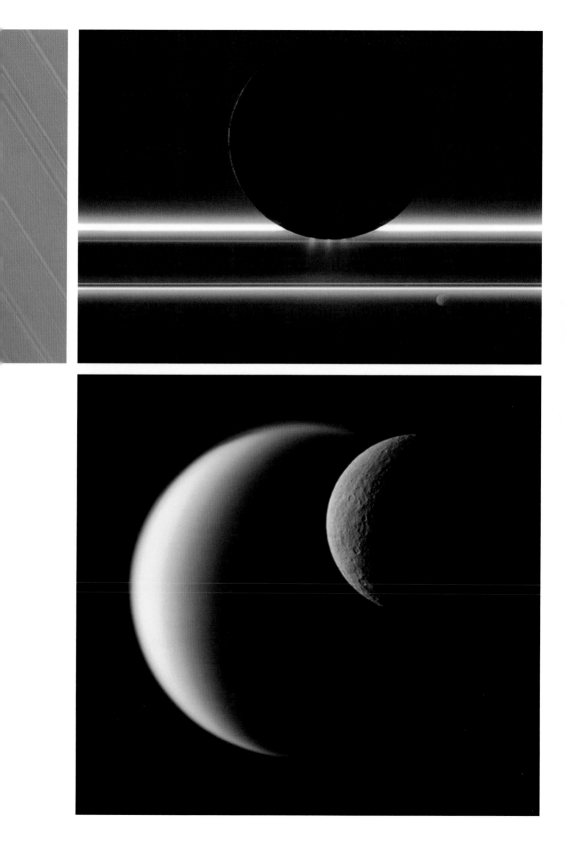

Above A blue bubble of foreboding. Technically known as a Wolf-Rayet nebula, this glowing sphere of dust and gas surrounds the unstable star that has ejected it. At a distance of 30 000 light years from our Solar System, the star will eventually end its life in a spectacular supernova explosion.

ESA/Hubble and NASA

Above A twisted maze of gas and dust surrounds the unstable star V838 Monocerotis. In January 2002, a brilliant outburst from the star created an expanding shell of light that now illuminates the material around it in a spectacular light echo, captured here by the Hubble Telescope.

NASA, ESA and H Bond (STScI)

Left Undisputed pearl of the northern sky, the nearby galaxy Messier 81. Its serenely beautiful spiral arms originate in a disturbance known as a density wave passing through the underlying gas, which triggers the formation of hot blue stars.

NASA, ESA and the Hubble Heritage Team (STScI/AURA)

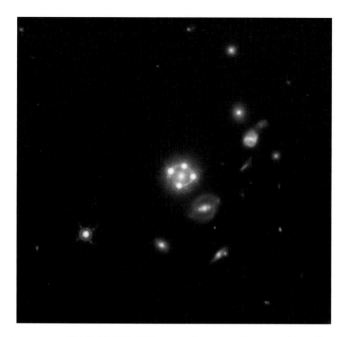

At the centre of this Hubble Telescope image, a distant galaxy distorts the space around it to act like a lens, embellishing it with four separate images of an even more distant – but much brighter – quasar. Such optical illusions are predicted by Einstein's General Theory of Relativity.

ESA/Hubble, NASA, Suyu et al

In this image, the lens-like effect of a cluster of galaxies has distorted a distant background galaxy into an incomplete circle known as an Einstein ring. The base of an ordinary wineglass provides an analogue of this gravitational distortion.

ESA/Hubble and NASA

This 'cosmic wallpaper' is the backdrop to everything we can see in space. Covering the whole sky, it is the ancient flash of the Big Bang, stretched into microwaves by the Universe's expansion during its 13.8-billion-year journey to our telescopes. The colouring represents tiny variations in temperature, which seeded today's galaxies.

ESA/Planck Collaboration

Despite their slightly comical demeanour, these two gentlemen revolutionised 19th century physics by unambiguously identifying the chemical elements in the Sun. Gustav Kirchhoff (left) and Robert Bunsen (right) collaborated at the University of Heidelberg in the 1850s and 60s, pioneering the modern science of spectroscopy.

Edgar Fahs Smith Image Collection, University of Pennsylvania

If ever a single image proclaimed the staggering complexity of the
Universe, this is it. The rich galaxy cluster Abell 370 is 4 billion light
years distant, but the mirage-like effect of its gravity reveals the
distorted images of galaxies more than three times further away.

NASA, ESA Hubble, HST Frontier Fields

like earthquakes, volcanic eruptions or asteroid impacts. The end product is a loss of genetic diversity, which, if the population survives, only recovers slowly over time. The loss is then reflected in the genetic make-up of the survivors' descendants.

Several studies over the past two decades have attempted to investigate such possibilities in relation to humans, and, in the mid-1990s, a theory emerged suggesting that around 70 000 years ago, the population was decimated to only a few tens of thousands of individuals worldwide. This is well into the era of modern humans, of course, and whatever caused the bottleneck may also have precipitated the decline of the Neanderthals who shared the planet (and quite often their genetic material) with *Homo sapiens* until about 40 000 years ago.

Such a population bottleneck would have required a catastrophic event that was global in scale, and the culprit is usually assumed to have been an eruption of the Toba supervolcano in Sumatra, which has been reliably dated at 74 000 years ago. This well-studied event is believed to have been the biggest volcanic eruption on Earth within the past million years, and ranks among the largest for which geological evidence remains. It deposited ash over much of southern Asia and the nearby seas, and produced a global cooling of up to 15 °C for at least a few years – a 'volcanic winter'.

Research published in 2018 by two international groups of scientists working independently casts doubt on this neat and tidy theory, however. Despite the fact that yes, thousands of cubic kilometres of ash were lofted into the atmosphere, and there is evidence of population bottlenecks among some animal species at around the same time, studies on a number of fronts indicate that human populations thrived in the aftermath of the eruption. Archaeological artefacts in southern India were as numerous above the ash layer as below it, for example. Doubt has also been cast on whether the genetic evidence supports a sudden population decline, or something that was spread over a longer period. This research clearly has a long way to go before definitive statements can be made. But you have to admit, it's thought provoking.

Thinking about escape routes, the word 'colonisation' is the one we most often hear applied to Mars. And that word not only implies settlement, but also control. Control over whoever (or, more likely, in an interplanetary context, whatever) is there already. Back in 2016, a keynote address by Elon Musk, a space entrepreneur whose work I usually admire, was littered with phrases about a Mars colonial fleet, a self-sustaining Martian city, and rapidly colonising the planet, hundreds of individuals at a time. Musk's vision includes cargo missions to Mars in the early 2020s, with the first humans arriving a few years later. He foresees a population measured in millions being established over the next 40 to 100 years.

That speech has been updated a couple of times since, but I'm afraid it still doesn't cheer me up. Talk of colonial fleets in the 2020s is bound to raise false expectations among an already escapist public. Such ideas also legitimise our lethargy when it comes to making our own planet a more sustainable and secure world. Why bother, if we're all off to Mars?

THERE'S MUCH TO ADMIRE IN THE WORK OF MUSK AND his SpaceX company. And I'm not a Luddite when it comes to spaceflight; I've been an unashamed enthusiast since childhood. It makes sense to conceive of humans as an interplanetary species. I wholeheartedly agree – humans *must* set foot on Mars.

But while Musk's populate-Mars-at-all-costs message seeks to inspire, it is flawed. The first problem is that the marketing of the idea is overly optimistic. Despite a fine track record within the private sector, and very considerable resources, the technology simply doesn't yet exist to take large numbers of humans on an interplanetary journey. Musk's *Starship* (formerly known as the *Big Falcon Rocket*) is currently under development, along with the *Falcon Super Heavy* launch vehicle that will boost it into space from the Earth's surface. With a previously declared goal of 100 Mars passengers per flight, this is huge engineering. And, due credit to Musk, he has maintained a policy of openly presenting the technical challenges (often via social media) and the 'very exciting' and 'delightfully counter-intuitive' engineering solutions that SpaceX proposes. But space engineering experts are not a little sceptical – although they acknowledge that Musk's reusable boosters are a game-changer. And it's clear that some barriers – like the radiation hazard en route to Mars – don't have a solution anywhere in sight.

Plans for sustaining the Mars colonists when they get there are likewise vague. Bioregenerative life support is touted as the gold standard in a hostile environment – life support in which plants and animals recycle human waste products (including exhaled breath as well as the stuff you're thinking of) for reuse. But scientists are still a long way from perfecting it. Experiments conducted on Earth over many years have demonstrated that true sustainability is extremely difficult to achieve. The *Biosphere-2* facility in Arizona required external supplies to keep its eight occupants alive during the 1990s, and is now used for more modest scientific experiments. A Chinese experiment, *Yuegong-1* ('Lunar Palace 1'), begun in 2014, seems to have been more successful. But it remains the case that all current human spaceflight is sustained by conventional life support systems, in which waste is discarded. It was *Apollo 9* astronaut Rusty Schweickart who observed that 'the most beautiful sight in orbit … is a urine dump at sunset', a sentiment that has been echoed by other astronauts at the sight of millions of snap-frozen ice crystals sparkling in the sunlight.

Perhaps Musk was attempting to up the ante when, in 2018, he declared that it was important to colonise Mars to preserve our species in the event of a third world war. 'If there's a third world war we want to make sure there's enough of a seed of human civilisation somewhere else to bring it back and shorten the length of the dark ages.' Quite so, but it doesn't have to be Mars. And how much better would it be to work on preventing that war in the first place?

More modest than SpaceX's ambitions, but considerably more loopy, were the plans of the Dutch colonisation project Mars One. Following its foundation in 2012, Mars One reset its goals several times, as the economics of its funding model were repeatedly demonstrated to be flawed. Originally the organisers planned to fund it with reality TV broadcasts, but the project grossly overestimated the likely revenue – and hugely underestimated the likely cost of flying humans to Mars. The intended innovation was to fund only one-way flights, so the would-be colonists would never come back home. On ethical grounds alone, it's doubtful that any launch agency would sanction that scenario. After several bail-outs, the organisation was liquidated early in 2019.

But the sad part is that tens of thousands of individuals applied to be among the 'lucky' colonists, believing that a life on Mars was their destiny. Something like 100 Mars One candidates progressed from the original intake of a couple of thousand applicants (not 200 000, as claimed by Mars One) to be shortlisted for flights. And the media delighted in interviewing such enthusiastic one-way trippers.

Because of the technological hurdles in sending humans to Mars, the overwhelming likelihood is that the first humans to set foot on the planet will be from one or more of the world's national space agencies, with the private sector being contracted to provide the hardware rather than supply the astronauts. And it's unlikely to happen before the mid-2030s. I hope I'm still around to see it.

MORE FUNDAMENTAL THAN ALL THE ENGINEERING problems, though, is the issue of colonisation. As we've seen on Earth, this has terrible consequences for indigenous populations. And while, as yet, we know of nowhere beyond our own planet where life exists, the fact is that we don't know whether Mars is as empty as it looks. NASA's *Curiosity* rover has already demonstrated that the planet was once habitable, and it hasn't ruled out all the possibilities yet.

Any ancient inhabitants might still be there. Where, for example, do the methane spikes observed today in the Martian atmosphere come from? Methane has been detected both from space and from the ground, and is significant because, unless it's replenished, it's quickly dissociated into its constituent carbon and hydrogen atoms by sunlight. Something must be replacing it. Living organisms (which are the source of most of Earth's atmospheric methane)? Or is it residual volcanic activity? Not quite as exciting, but still an intriguing possibility. The European Space Agency's *ExoMars Trace Gas Orbiter* is sampling Mars' atmosphere as I write these words, in order to find out.

Intriguing, too, is the mid-2018 discovery of liquid water beneath Mars' southern ice-cap. High-intensity radar reflections spotted in polar data from ESA's orbiting *Mars Express* spacecraft can only have come from a large body of liquid water near the base of the 1.5-kilometre-thick ice-cap. It is probably kept liquid at its estimated temperature of −68 °C by dissolved mineral salts, whose presence on Mars is already known from surface landers. This is a discovery of extraordinary significance, and is bound to heighten speculation about the presence of living organisms on the red planet. Caution needs to be exercised, however, as the concentration of salts needed to keep the water liquid could be fatal for any microbial life similar to Earth's. With no immediate means of sampling the water, the jury remains out as to the possibility of the newly discovered lake harbouring life.

In 2020, however, both NASA and the European Space Agency will fly new rovers to Mars with the explicit purpose of seeking evidence of past or present life there. NASA's *Mars2020* and ESA's *ExoMars* landers will target environments on Mars that are of astrobiological interest – although the southern ice-cap has already been ruled out. Several other organisations are contributing to these missions, including, in the case of *ExoMars*, the Russian space agency, Roscosmos.

My guess is that within the next few years, we'll see firm geological evidence of past life on Mars. That was already hinted at in 2016 when data from NASA's *Spirit* rover revealed opaline silica outcrops in Gusev Crater that closely mimic biologically altered deposits at El Tatio in northern Chile. The discovery of past life will hasten the search for current life, and facilitate the continuing exploration of the planet. When that exploration involves humans, it will be a fantastic breakthrough. But, as far as possible, it should be done within stringent rules – taking into account a possible relaxation of those rules to accommodate human exploration. The rules are aimed at protecting possible indigenous biospheres of other planets. What right have we to interfere in an evolutionary process that might, in a few billion years, see intelligent Martians with their own technological capabilities?

The thought of millions of colonists tramping over the pristine surface of Mars is as unpalatable to me as the idea of setting up sprawling cities in Antarctica – which, by the way, would be far easier than setting them up on Mars. And there are other possibilities. Jeff Bezos of Blue Origin has outlined a long-term vision that involves artificial megastructures in space, built using materials sourced from asteroids. With gravity provided by slow rotation, and an environment that Bezos likens to 'Maui on its best day all year long', his thinking aligns much more closely with mine. May I suggest, therefore, that we leave the exploration of Mars to the explorers – the specialist few (as, indeed, we do in Antarctica) – while the rest of us cheer them on from the sidelines? And, of course, while we get on with making our own planet a more sustainable and secure world. Now that's the real challenge.

CHAPTER 12

RINGING IN THE CHANGES: THE VANISHING RINGS OF SATURN

The planet Saturn is a close second to the Moon in its appeal to budding astronomers, and guaranteed to captivate rookie skywatchers using small telescopes. When viewed with a magnification of 30× or more, the ringed giant takes on a surreal yet strangely familiar appearance that nearly always causes gasps of delight – especially when seen for the first time. Can that possibly be real, they ask? Oh, yes it can.

To the first person who *ever* saw it through a telescope, however, it was a source of frustration. Perhaps the fact that Saturn was then the most remote known planet compounded the mystery that Galileo Galilei encountered when he began observing it. He could tell it was different from the other worlds he'd studied with his home-made telescope in the northern spring and summer of 1610, but couldn't work out what was going on. Venus, Mars and Jupiter all showed clear discs of light – with Moon-like phases in Venus's case – but Saturn's disc seemed to have appendages. In a couple of throwaway paragraphs on page 25 of his *Letters on Sunspots* of 1613, he comments that the planet looks like 'three [stars] together, which almost touch each other'. He wondered if they were handles, or even ears? The mystery deepened when Galileo observed the planet again in 1612, and found it had turned into a perfectly respectable unadorned disc of light. Had Saturn devoured his children, he wondered, as the god Saturn was wont to do?

We now know that Saturn's rings were invisible in 1612 because its 29-year orbit had taken it to a position where they were edge-on to the inner Solar System. The tilt of the rings to the plane of Saturn's orbit is 26.7 degrees, rendering them visible throughout most of the orbit – except during the two equinoxes of the Saturnian year, when the Sun is overhead on the planet's equator. From our vantage point, they seem to vanish because they have hardly any thickness. Being edge-on to the Sun, too, they cast no tell-tale shadow on the planet. Not long after each equinox, the rings become visible again, and by 1616 Galileo was noting that the planet's appendages had taken on a larger and more elliptical appearance. No wonder he was baffled.

It took almost four decades of frustrated speculation by a succession of 17th-century astronomers before the Dutch nobleman and scientist, Christiaan Huygens, worked out that Saturn must be surrounded by 'a thin, flat ring, nowhere touching, and inclined to the ecliptic'. He made this discovery in 1655, at the grand old age of 26. Huygens' book *Systema Saturnium* (1659) expounded the idea, including the correct explanation for the rings becoming invisible every 14 years or so. It was the start of humankind's love affair with Saturn, and very soon, other astronomers were adding their ten penn'orth on the nature of the rings. Perhaps because he was better known as a poet than an astronomer, the one person who guessed the right answer in 1660 was ignored for almost 200 years. Jean Chapelain suggested that the ring was composed of many small orbiting bodies, rather than solid material as Huygens had proposed. Others noted that there was not just one, but several rings surrounding the planet – some with narrow gaps between them.

It was the great 19th-century Scottish physicist, James Clerk Maxwell, who vindicated Chapelain. *On the Stability of the Motion of Saturn's Rings*, published in 1859, contained Maxwell's mathematical demonstration that solid rings couldn't withstand the pull of Saturn's gravity. They must be composed of an 'indefinite number of unconnected particles' in individual orbits around the planet. He went on to describe the rings as 'the most remarkable bodies in the heavens, except, perhaps … the spiral nebulae'. Maxwell's result was confirmed observationally in 1895 by American astronomer James Keeler. Cleverly observing the rings with a spectrograph – which can be used as a kind of celestial speedometer – Keeler showed that the inner edge rotates faster than the outer edge, which is the opposite of what would happen if the rings were solid.

Measurements made in 1970 (again using the diagnostic capabilities of a spectrograph) showed that the rings are made predominantly of ice – ordinary frozen water. We now know that the myriads of ice particles range in size from dust grains to boulders 10 metres or so across. It's the gravitational pull of the planet and its moons that hones them into a blade of material some 250 000 kilometres in diameter, but, astonishingly, less than 100 metres thick.

IN 1979, NASA'S *PIONEER 11* BECAME THE FIRST SPACECRAFT to study Saturn and its rings during a fleeting encounter with the planet. The trouble with spacecraft undertaking grand tours of the planets, as *Pioneer 11* was, is that they can't hang around. It sped past Saturn at a cool 32 kilometres per second. Among *Pioneer*'s discoveries were a previously unknown moon (Epimetheus – which it nearly collided with) and a narrow ring on the outer edge of the main ring. That main ring, by the way, is known as the A-ring, and the new one – being the sixth to be identified as a separate entity – became the F-ring. In reality, all the rings of Saturn are made up of finer 'ringlets', which appear to merge together when seen from a distance.

Pioneer 11 paved the way for *Voyager 1* and *Voyager 2*, which flew by Saturn in 1980 and 1981 respectively. Spoke-like features in the main ring system and braided strands in the F-ring were among their legacy discoveries, setting the scene for one of the most productive space projects ever – the incomparable *Cassini*. Launched in 1997, *Cassini* arrived at Saturn in July 2004, and provided extraordinary images and data on the planet, its rings and its moons for 13 years. You'll notice it makes several starring appearances in this book. Its demise, in September 2017, was one of the most poignant moments in the history of spaceflight, when the craft was intentionally flown into Saturn's outer atmosphere to vaporise and become part of the planet it had so generously laid open for humankind. Remarkably, new discoveries have continued to emerge from the copious data it provided, long after it has gone. And they include some surprising Saturnian secrets.

The question of why the planet has such magnificent rings, making it undoubtedly the pearl of the Solar System, has occupied the minds of scientists since Huygens' time. As has a related conundrum: were the rings formed with Saturn, or are they a more recent addition? While those questions haven't yet been completely answered, we now have a much better idea of how – and, in particular, when – the rings originated. This emerges from an understanding of their ultimate fate, because it turns out that the rings are not permanent.

RINGS ARE NOT ESPECIALLY RARE IN THE COSMOS. MANY celestial objects have them – including a clutch of minor Solar

System bodies. The icy asteroid–comet hybrids known as Centaurs, for example, include at least two with rings. (They're Centaurs because they, too, were hybrids: half man, half beast. Hey, who says astronomers are unimaginative?) The ring-bearers are Chariklo, whose rings were discovered in 2013, and Chiron, which astronomers have suspected to have rings since the 1990s. Both objects are around 200 kilometres in diameter and sit between the orbits of Saturn and Uranus. A much larger example is the dwarf planet Haumea, a football-shaped object around 2000 kilometres long that circulates beyond the orbit of Neptune. It takes a leisurely 284 years to travel once around the Sun, but in stark contrast, rotates on its axis in the breathtakingly short time of 3.9 hours – hence its elongated shape.

You might wonder how the rings of such distant worlds are discovered, given that they must be very faint. There is a standard procedure, which involves many observers stationed over a wide area of the Earth's surface. They're watching for the light of selected stars dimming as asteroids or dwarf planets pass in front of them. This involves carefully predicting the way such objects track, of course, but these calculations are the stock in trade of Solar System astronomers. By measuring how much the target star dims, and timing it carefully, a track of the asteroid's shadow can be plotted, cast in starlight on the Earth's surface. That allows the astronomers to determine its shape and size accurately, and to discover if it has rings or moons too. The passage of one celestial object in front of another one is known as an occultation, from the verb to occult, meaning to hide. The occultation method is widely used in the study of minor Solar System objects.

Stepping up to the other end of the size scale, all the giant planets of the Solar System have rings, but those of Jupiter, Uranus and Neptune are narrow and decidedly anaemic. None of them remotely approaches Saturn's for imposing breadth and brilliance.

There may, however, be a close link between them and the ultimate fate of Saturn's lavish adornments.

IT WAS THE TWO *VOYAGER* SPACECRAFT THAT FOUND THE first evidence of icy material from the rings raining onto the planet's cloud belts. It came from studies of changes in the electrical charge of Saturn's outer atmosphere, along with the discovery of some mysterious dark bands in the normally bright cloud belts of the planet's mid-latitudes. Add to this mix some brightness variations of the rings themselves, and you have what seems like a hotchpotch of unrelated effects. But, in 1986, a scientist by the name of Jack Connerney at NASA's Goddard Space Flight Center managed to draw these strands together, and infer that icy particles from the rings were acquiring an electrical charge from the Sun's radiation, and then being funnelled by the planet's immense magnetic field down towards the cloud belts. There, they rinsed away the haze that makes the clouds bright, resulting in dark bands.

Connerney made some estimates of the rate at which this 'ring rain' was draining material from the rings, obtaining figures in excess of one tonne per second. Late in 2018, this was confirmed by data obtained with one of the two 10-metre Keck telescopes at Mauna Kea Observatory in Hawaii. Astronomers there detected infrared radiation emitted when the ice particles spiral downwards along the magnetic field lines, dumping water into Saturn's middle latitudes north and south.

At around the same time, a large group of researchers from predominantly US universities published new results from the 22 audacious Grand Finale orbits of the *Cassini* mission in 2017. You might remember that with nothing to lose, the spacecraft was famously threaded between the rings and the top of Saturn's atmosphere. As it dived through the 2000-kilometre-wide gap,

Cassini detected a direct flow of ice from the inner edge of the rings to the planet's equator. It also measured the flow's composition, and found it to be unexpectedly rich in complex carbon-containing compounds, which amounted to some 37 per cent of the infalling material. Surprisingly, since water ice is by far the most abundant component of the rings, water itself comprised only 24 per cent, with the remainder being made up of methane, carbon monoxide and nitrogen.

Adding together these two processes of ice spiralling down along Saturn's magnetic field lines and ice falling directly towards the planet's equator allows scientists to establish that at least 10 tonnes are draining from the rings every second. Some measurements suggest the rate may be four times this amount. Rather alarmingly, even the more conservative estimate implies that the rings will vanish altogether in less than 100 million years. Better take a good look at them while they're still there.

It's also likely that this drainage is the reason Saturn's innermost rings – the D- and C-rings – are fainter than the others. It suggests the rings are leaking down to the surface from their inner edge, with material being pulled from successively further out in the ring system. It's not hard to imagine the final stages of the process producing something that looks a lot like the emaciated rings of the other gas giant planets. And that, in turn, suggests that for Jupiter, Uranus and Neptune, we've kind of missed them at their best.

AND SO TO THE $64 000 QUESTION – HOW OLD ARE SATURN'S rings? Their rapid decay hints that they are much younger than the planet itself, which was born some 4.6 billion years ago together with the Sun and the rest of the Solar System. This idea is supported by the fact that the rings are generally bright, suggesting they are largely uncontaminated by the dusty interplanetary debris that would otherwise have accumulated within them.

There's one further measurement that could clinch this suggestion and it too was made during the Grand Finale orbits. As *Cassini* plunged between the planet and its rings, it responded to the gravitational pull of both, allowing the rings to be weighed. In January 2019, the results of that measurement were announced, and it's a grand total of 15 thousand trillion tonnes (or, putting it a bit more scientifically, 1.5×10^{19} kilograms). Despite that imposing number, it's actually very small – about half the mass of the Antarctic ice sheet. Knowing the mass of the rings allows planetary scientists to model their age accurately, and the answer supports the idea of a recent origin. They now believe the rings are between 10 million and 100 million years old – the blink of an eye compared with the age of Saturn. It's extraordinary to think that when dinosaurs roamed Earth, Saturn was probably ringless.

And, finally, where did the rings of Saturn come from? It's now generally believed that this spectacular but temporary phenomenon was caused by one or more icy objects breaking up. Perhaps a large comet strayed too close to Saturn and was broken into myriads of icy fragments by the planet's pounding gravity. Or possibly there was a collision between two or more icy Saturnian moons. The mass of the rings is certainly comparable with that of some of the planet's smaller satellites. More than this we may never know – not, at least, until there is another Saturnian space mission to rival the marvel that was *Cassini*.

CHAPTER 13
STORMY WEATHER: WEIRD WORLDS OF THE SATURNIAN SYSTEM

Cassini's discoveries about Saturn's rings are undeniably sensational, but I'd venture to suggest that its findings concerning the planet itself and its extensive systems of moons are even more amazing. During the spacecraft's 13-year sojourn, it completed no fewer than 294 orbits of Saturn in four distinct phases. First came the Prime Mission (2004–2008), which gave scientists their first up-close-and-personal view of the Saturnian system. That was extended to become the Equinox Mission (2008–2010) covering Saturn's equinox of 11 August 2009, when the Sun illuminated the rings exactly edge-on. (What does edge-on illumination do? It throws any 'vertical' structures in the rings into relief by virtue of the shadows they cast, revealing, for example, graceful waves in the ring system produced by the gravity of small moons orbiting within it.)

And, with the spacecraft still operating flawlessly, the project was extended again to cover the northern summer solstice on 25 May 2017. This became the Solstice Mission (2010–2017), during which *Cassini* gained spectacular views of the northern polar regions of the planet and some of its moons bathed in summer sunlight. Finally, with *Cassini* running out of fuel for orbital manoeuvring, caution was thrown to the winds with the 22 Grand Finale Mission orbits. For five months, we held our collective breath as the spacecraft repeatedly passed – surprisingly unscathed – between the planet and its rings, ending its epic mission on 15 September 2017.

The *Cassini* mission's abundant discoveries were down to an army of scientists taking the raw data and turning it into new knowledge. They number in the hundreds, and are based at universities and scientific institutions all over the world, but I'll mention two of the key figures, without whom the project might not have been anywhere near as successful. First was Linda Spilker, *Cassini* project scientist at NASA's Jet Propulsion Laboratory (JPL) in Pasadena, California. In other words, the Head Honcho. Linda cut her teeth in planetary science with the *Voyager* missions, which she joined when the two spacecraft were launched to the outer Solar System in 1977.

I had the honour of meeting Linda in Pasadena when I was visiting the United States for the Great American Total Solar Eclipse in August 2017, less than a month before *Cassini*'s final dive into Saturn's atmosphere. After we'd chatted about the science, I asked her whether she'd be sad at the spacecraft's demise, having spent much of her career preparing for the mission, cheering it on during its seven-year voyage to Saturn, and supervising its science program. 'No,' she replied. 'I'm already focused on what comes next.' But when I watched the live broadcast from JPL at the mission's end, along with millions of others around the world, I couldn't miss the handkerchief she had at the ready. 'It's like losing an old friend,' she told the media afterwards, and I'm not surprised. I felt like that, too, and I was only an enthusiastic bystander.

The other person I'd like to single out is Carolyn Porco, who led *Cassini*'s Imaging Science Team throughout the mission's observational phase. Carolyn is much more than an expert scientist, however. She also has the eye of an artist, and many of the half-million or so images obtained by the spacecraft are breathtaking in their beauty. And that's not just because of the extraordinary subject matter, but because of their composition, detail, colour, lighting and object juxtaposition – all the ingredients of an awesome picture. Of course, the images are also scientifically valuable, greatly enhancing our understanding of the Saturnian system.

Saturn's atmosphere, for example, beguiles planetary scientists with its complex structure. The planet is a 'gas giant', a world with no detectable solid surface, shrouded in dense clouds. Unlike Jupiter, whose cloud belts are visible from Earth even in small telescopes, Saturn has rather subtle markings, because its 'weather' occurs at lower atmospheric levels due to the colder temperature at its greater distance from the Sun.

However, a few eagle-eyed amateur astronomers equipped with state-of-the-art electronic detectors on their hobby telescopes were able to pinpoint storms in Saturn's cloud belts as they developed, alerting *Cassini* mission scientists to unusual Saturnian meteorology for immediate follow-up. Probably the best known is Trevor Barry of Broken Hill in outback New South Wales, a retired miner I had the good fortune to get to know almost 20 years ago. Inspired by the stars, Trevor embarked on an astronomy degree at Swinburne University after his retirement, and wound up with the university's Award for Excellence as the top graduate of his year. He has an enviable track record of Saturnian storm-spotting, a talent that quickly convinced *Cassini* scientists of the value of working with this unassuming and down-to-Earth astronomer. One particularly active mid-latitude storm turned out to be a record-breaker, and Trevor's scientific rags-to-riches story led him not only to collaborate with Carolyn Porco and

HOW DOES A HEXAGON...

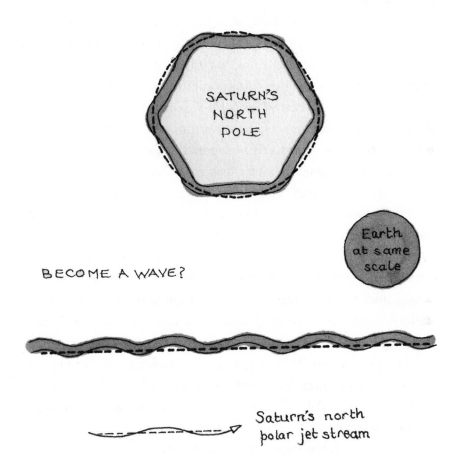

SATURN'S
NORTH
POLE

Earth
at same
scale

BECOME A WAVE?

Saturn's north
polar jet stream

While Saturn's northern polar hexagon looks almost artificial, this diagram shows how it is formed by a six-fold standing-wave pattern in the planet's polar jet stream. Atmospheric vortices near the 'points' of the hexagon keep it stable.

Author

her team in Pasadena and visit the high-altitude observatories in Hawaii, but also to star in his own feature segment on Australian national television.

As Saturn moved on from its equinox and springtime sunlight began to illuminate the planet's arctic region, *Cassini*'s cameras revealed a ferocious hurricane that rages around its north pole. With wind speeds around 500 kilometres per hour, and an 'eye' that spans 2000 kilometres, it's truly a giant among storms. But it sits in the exact centre of something more extraordinary still – a belt of clouds that is perfectly hexagonal in shape.

The six straight sides of this strange geometric pattern are each bigger than Earth. To be honest, in pictures it looks more like something you'd take a spanner to than a natural phenomenon, but it's now understood to be due to the planet's polar jet stream. Like Earth's own jet streams, it meanders from side to side as it circulates, but unlike ours, it is not disturbed by continents and oceans underneath. Thus, it has settled into a stable wave pattern with six 'peaks' and 'troughs' around its circumference – a circular standing wave forming a perfect hexagon.

Despite the consistent shape, it does display changes in its colouring. Images taken in November 2012, early in the Saturnian northern spring, showed that the interior of the hexagon is bluish in hue, and quite dark. Four years later, the blue had turned to a rich golden colour, similar to the rest of the planet. Scientists think this gold tint comes about because of sunlight-induced chemical reactions in Saturn's atmosphere, producing more suspended particles (aerosols) and leading to a greater level of haze.

IF STORMY WEATHER IN SATURN'S ATMOSPHERE IS UNEXpected, one doesn't have to look far away to discover even more bizarre conditions. Many of the planet's moons are strange worlds, but its largest satellite, Titan, is by far the strangest. At 5150 kilometres in diameter, it is the second biggest moon in the Solar System after Jupiter's Ganymede – larger than the planet Mercury, and half as big again as our own Moon. It was discovered on 25 March 1655 by the same chap who figured out that Saturn had rings around it – the great Dutch astronomer and mathematician, Christiaan Huygens.

Titan takes 15 days and 22 hours to travel around Saturn, and the same length of time to rotate on its axis. Thus, like our own Moon, it always keeps the same face towards its parent planet – but there the similarity ends. Titan is the only moon in the Solar System to have a thick atmosphere, which stabilises its surface temperature at around –180 °C. And, whereas our Moon's surface is solid rock overlaid with thin soil, Titan's surface is rock-hard water-ice over which erosion processes have created a 'sand' of ice crystals and solidified hydrocarbons. This material forms long, wind-blown dunes in Titan's equatorial regions.

There's considerable evidence that Titan's icy surface forms a shell that floats above a global ocean of liquid water and ammonia, kept warm by nuclear processes in Titan's rocky core. We know the ice-shell rotates independently of the core, because the longitudes of geographical features on its surface display a small backwards and forwards motion as Titan orbits Saturn. And, as if that wasn't weird enough, it's thought that Titan has a number of freezing volcanoes, spewing out a magma composed of slushy water and ammonia. Only one has been confirmed, however.

Most city-dwellers in sunny climates are familiar with that orange haze that sometimes develops in the atmosphere on windless summer days. My home town of Sydney is famous for it, sitting as it does in a basin between mountains and ocean. The haze is a photochemical smog, caused by the action of the Sun's ultra-violet rays on hydrocarbons primarily from vehicle exhausts. There are no vehicles on Titan (apart from one now-defunct robotic lander), but its atmosphere has a similar composition – mostly nitrogen, but laced with a brew of methane and other hydrocarbons that cause the opaque orange haze. Thus it's difficult to map the surface of Titan, even from space.

Despite the hazy atmosphere, we know that Titan has a weather cycle of evaporation and rainfall, similar to that on Earth. However, the moisture in its atmosphere is not water vapour (which would be frozen), but a mix of hydrocarbons that are best thought of as liquid natural gas – ethane, methane and other compounds. Indeed, clouds of this ethane–methane mix usually cover a small percentage of Titan's surface, and when conditions are right, rain falls from them.

IT WAS DATA FROM THE FLY-BYS OF THE TWO NASA *VOYAGER* probes in the early 1980s that suggested the possibility of hydrocarbon seas on Titan. By the mid-1990s, American astronomer Carl Sagan and others had suggested there might be ocean-sized bodies of liquid methane on the surface, based on ground-based radar data. Once *Cassini* arrived in orbit around Saturn in 2004, the hope was that large bodies of liquid would be very quickly detected. The spacecraft carried a small lander called *Huygens*, and when that touched down on Titan's surface on 14 January 2005, some expected it to splash down in an ocean. It didn't, but images sent back during its parachute descent revealed drainage channels leading to what could be a shoreline. Tantalising stuff.

By 2007, however, scientists believed they had definitive evidence of lakes filled with methane, which came from smog-penetrating radar aboard the *Cassini* orbiter. These lakes were mostly near Titan's north and south poles, and their existence has now been confirmed beyond doubt by radar and infrared mapping. They pool in basins in the 'bedrock' of ice on Titan's surface, and are the only stable bodies of liquid known anywhere in the Universe, other than those on Earth. The seas and lakes dominate in Titan's northern arctic, although there are a few in the south. They are large – comparable in area with North America's Great Lakes in the case of the three biggest, which are designated as maria, or seas. Titan's largest sea, Kraken Mare, is about three times larger than Lake Michigan-Huron, which, with a surface area of 117 300 square kilometres, is the biggest freshwater lake on our own planet.

Some 30 smaller lakes, ranging from a few kilometres in length up to a couple of hundred, have also been identified. All these polar seas and lakes appear to be fed by methane rainfall (via river-like features), but there are a few lakes in Titan's equatorial region that are probably fed by springs from a methane and ethane 'water table' in places where the ice bedrock is porous.

The radar equipment carried aboard *Cassini* is also capable of measuring the depth of Titan's lakes and seas. Average depths vary from two or three metres for the smallest lakes to tens of metres for the seas, with a maximum depth of more than 200 metres (the limit of measurement) for Ligeia Mare, Titan's second biggest sea. It's also possible to use radar to detect the average wave height on the lakes and seas, and the measurements that have been made show very small waves – around a few millimetres in height. This suggests either that surface winds are very low, or the liquid in the lakes is oily – or perhaps both.

Although we now know a lot about Titan's lakes and seas, many tantalising questions remain. One concerns temporary surface features that have been observed in the three large seas – Kraken Mare, Ligeia Mare and Punga Mare. They look like bright patches that seem to come and go. But bearing in mind that these features are detected by radar reflection (where a bright signal indicates a rough surface), some scientists have attributed them to surface ripples on the seas, whipped up by light winds. An alternative hypothesis is that they are methane 'icebergs', which form on or near the surface, and then sink from view as the conditions change.

Also hypothesised is the prospect that cyclones occur over the three large seas, with some predicting that Titan's summer weather could produce the necessary conditions. Perhaps *Cassini*'s demise came too early in the summer, for none were actually observed. Similarly tumultuous is the 'Throat of Kraken', a narrow neck of liquid in Kraken Mare that is expected to generate strong currents, and perhaps even whirlpools, at certain seasons of Titan's 29-year journey around the Sun.

TITAN IS, INDEED, A STRANGE WORLD, BUT IT MAY HOLD even more dramatic secrets. With suspicions of a rich organic (carbon-containing) chemistry on the surface borne out by observations already made, some scientists believe this frigid place is an analogue for the early Earth, with an atmosphere similar to that of our own planet before life evolved. Others go further, suggesting that there could already be life-forms thriving in the hydrocarbon lakes. They would be quite different from the water-based life we see on our own planet, using liquid methane as their working fluid, breathing hydrogen and feeding on acetylene. Tantalisingly, both these chemicals are depleted at low levels in Titan's atmosphere.

This is by no means evidence for life on Titan – there are abiotic processes that could equally well produce the same effect. But it is a hint that there may just be life in the Solar System so radically different from life on Earth that it could only have formed independently. And, should such a 'second Genesis' be proved correct, it would suggest that life might well be widespread throughout the Universe.

With that intriguing thought in mind, a number of space-craft have been proposed to further explore Titan, with particular interest in the seas and lakes. They range from a balloon-borne robot floating in Titan's atmosphere to a robotic submarine to explore the seas. To date, just one has been funded – NASA's *Dragonfly* drone rotocopter, announced in June 2019 and sched-uled for launch in 2026. It's likely others will follow.

CHAPTER 14

STALKING AN INVISIBLE PLANET: THE SEARCH FOR PLANET NINE

Early in 2016, the world's science media ran wild with a story about a ninth planet orbiting the Sun – an unbelievably remote planet, with perhaps ten times the mass of Earth and up to four times its diameter.

Now, of course, some astronomers maintain there's already a ninth planet in the Solar System. They reckon it was discovered in 1930, and is called Pluto. Back in 2006, they'd been seriously annoyed when the governing body of astronomy (the International Astronomical Union – or 'übernerds', as the unkinder news outlets put it) finally got around to defining what constitutes a planet. The infamous result of that definition was that Pluto didn't make the cut, because it was not the gravitationally dominant body in its region of the Solar System – as you now need to be to be counted as a planet.

But back to the hypothetical ninth planet. How can we see it? Well, actually, we can't. The 'discovery' is an inference based on the movements of celestial objects that we can see – members of a family of icy asteroids way out in the Solar System, far beyond the orbit of Neptune. These so-called extreme trans-Neptunian objects (eTNOs) lie well beyond the ring of icy asteroids known as

142

the Kuiper Belt – whose best-known member is the dwarf planet Pluto. And it's a mathematical study of their orbits that provides the smoking gun for the postulated ninth planet.

Pluto itself doesn't feature in the quest for this so-called 'Planet Nine'. The search centres around a clutch of eTNOs – smaller and much more remote objects, of which the largest, Sedna, is about 1000 kilometres in diameter, roughly half the size of Pluto. Sedna is currently nearly three times further from the Sun than Pluto – or about 90 times further than Earth is from the Sun. I say 'currently' because, like most of these distant icy asteroids, Sedna has a highly elongated orbit, whose furthest reaches are more than ten times its present distance. As you might expect, Sedna's progress along this trajectory is pretty leisurely, taking about 11 400 years to complete a full circuit.

So how do these distant objects tell us there's a burly planet hiding out there? It was the discovery of Sedna in 2004 that began the story, but a decade later, US astronomers Chad Trujillo and Scott Sheppard pointed out a curious anomaly linking the orbits of Sedna and several smaller eTNOs. Their stretched-out orbits line up in a way that's quite different from the random alignments that would be expected from our present knowledge of the Solar System. Trujillo and Sheppard speculated that perhaps a massive and as-yet unknown planet was shepherding the orbits into alignment. It was further work at the prestigious California Institute of Technology – Caltech – by astronomers Mike Brown and Konstantin Batygin that caused the media's breathless enthusiasm in 2016.

Their calculations not only yielded a probable mass for the unseen Planet Nine, but also its likely orbit, and also hinted that some other minor weirdnesses of the Solar System could be explained by its presence. The planet Brown and Batygin envisaged is never nearer to the Sun than 200 times the Earth–Sun distance, and its highly elongated orbit might take it out six times further still. Its 'year' is estimated to be between 10 000 and 20 000 Earth years. Such was the excitement Brown and Batygin stirred up that their work initiated a search for the elusive planet, which, as I write these words, has not yet been located.

BUT IF THIS HYPOTHESISED PLANET IS REALLY FOUR TIMES the diameter of Earth, why haven't we found it yet? It could be that it's been camouflaged against a matching background. Its most probable location is somewhere in the furthest reaches of its elongated orbit, since that's where anything in such an orbit moves most slowly and thus spends most of its time. As we have seen, its distance in this position is likely to be as much as 1200 times the Earth's distance from the Sun – considerably more than remote Sedna's. It means that despite its size, it will be very faint, and point-like in appearance rather than disc-like. And, by a truly unlucky accident, its probable direction is in the most crowded part of the sky, the Milky Way. So imagine trying to locate a target that looks exactly like millions of stars around it, and is distinguished from them only by the fact that it is moving excruciatingly slowly across the sky. No wonder Planet Nine hasn't turned up yet.

THIS THEME OF HYPOTHETICAL PLANETS INFLUENCING the orbits of known celestial objects has some interesting antecedents in astronomical history. The best known was, indeed, a triumph for mathematical discovery, and resulted in astronomers extending the Solar System's inventory from seven planets to eight.

The story starts with the great 17th-century scientist Isaac Newton. Once he had published his Theory of Universal

Gravitation in 1687, there was no holding back the astronomers of the day. Very quickly, they found that the new theory perfectly accounted for the motions of all the objects in the Solar System out to its known boundary. In Newton's time, that was represented by the planet Saturn, but in 1781, William Herschel serendipitously discovered the planet eventually named Uranus after the god of the sky, a suggestion by the German astronomer Johann Elert Bode. (In fact both 'Herschel' and 'The Georgian Star' had been suggested as names for it, either of which would have spared us all a lot of crass jokes.)

Careful studies of Uranus's orbit in the immediate aftermath of its discovery revealed that something seemed to be pulling it slightly out of position. During the first half of the 19th century, the best mathematicians of the day attempted to deduce what that might be – most notably John Couch Adams in Cambridge and Urbain Jean-Joseph Le Verrier in Paris, who were working independently. Adams' 1845 prediction of a new planet was greeted unenthusiastically by the director of the Cambridge Observatory, James Challis, who declined to look for it. Apparently, he believed the predicted position was too inaccurate, and the deferential Adams was unwilling to risk his career by pushing too hard. But in 1846, Le Verrier published a more accurate prediction, which was in agreement with Adams', belatedly spurring Challis and the Astronomer Royal, Sir George Airy, to initiate a proper search.

Meanwhile, Le Verrier, having also failed to raise enthusiasm for a search in France, sent his prediction to Johann Gottfried Galle at the Berlin Observatory. Armed with the accurate position, it took Galle only an hour on 24 September 1846 to find the planet we now call Neptune. In the aftermath of the discovery, there was much controversy in English and French astronomical circles about who had priority in the finding, but the words of the Paris

Observatory director, François Arago, sum up the situation well. 'Le Verrier,' he wrote succinctly, 'has discovered a planet with the point of his pen.'

This triumph of gravitational theory was followed by an event that falls somewhere between scientific hubris and the recognition that no theory is guaranteed to be complete, no matter how distinguished its author is. Once again, the star of the show was the redoubtable Urbain Le Verrier. In 1859, still flushed with his success in predicting the existence of Neptune, Le Verrier returned to another problem that had bugged astronomers for nearly two decades – some unexplained behaviour in the orbit of Mercury. Once again, he used the mathematics of Newton's theory to predict that a small planet must exist within Mercury's orbit – a planet big enough to change Mercury's motion by its gravitational pull, but small enough to hide in the glare of the Sun. He suggested that the predicted planet be named 'Vulcan'.

On the publication of Le Verrier's prediction, astronomers all over the world carried out searches for the new planet. Since he'd been right about Neptune, everyone assumed he'd be right about Vulcan, too. Several astronomers throughout the later 19th century reported sightings of the elusive object, and Le Verrier went to his grave in 1877 firmly convinced of its existence. But it was never confirmed, and interest gradually faded in the wake of his death.

Then, in 1915, Albert Einstein produced his new theory of gravitation, known as the General Theory of Relativity. Today, it is the bedrock on which modern astrophysics is built. Crucially, it is in strong gravitational fields where its predictions differ most from Newton's. And where might you find a strong gravitational field in the Solar System? Close to the Sun, of course.

Just before he published his new theory, Einstein applied it to the orbit of Mercury and discovered that it exactly explained the observed anomalies that had led to the idea of Vulcan. He was ecstatic. 'For a few days,' he wrote, 'I was beside myself with joyous excitement.' And no wonder. At last, the myth of Vulcan had been laid to rest.

CURIOUSLY, HOWEVER, THE VULCAN MYTH WAS ALREADY being echoed in growing excitement about the possibility of another undiscovered planet – this time one beyond the orbit of Neptune. Since the closing years of the 19th century, astronomers had suggested that observed irregularities in the orbits of both Uranus and Neptune were the result of gravitational disturbance by a ninth planet.

An intensive search for this so-called 'Planet X' was started by Percival Lowell, an astronomer of independent means who had founded an observatory in Flagstaff, Arizona in 1894. Lowell pursued the quest until his death in 1916. After a hiatus resulting from a contested will, the search was resumed at Flagstaff in 1929 by Clyde Tombaugh, a young Illinois farmer with a passion for astronomy. On 18 February 1930, Tombaugh found a distant, slowly moving object roughly in the position that had been predicted by Lowell. It was quickly named Pluto, courtesy of a suggestion by an 11-year-old schoolgirl in Oxford. This was the remarkable Venetia Burney, whose grandfather passed on her suggestion to Oxford's professor of astronomy, Herbert Hall Turner – who, in turn, cabled it to Flagstaff. By the time of her death in 2009 at the age of 90, Venetia had witnessed not only the launch of a spacecraft to Pluto, but also its reclassification as a dwarf planet, both of which happened in 2006.

The discovery was greeted with universal enthusiasm – here was the evidence that the observed irregularities in the orbits of the outer planets were due to gravitational perturbations, in a further triumph of Newtonian gravity. Pluto's great distance made diameter measurements very difficult, but it was assumed to be a large planet – perhaps bigger than Earth. As the 20th century progressed, however – and astronomical equipment improved – estimates of Pluto's diameter gradually grew smaller. We now know it's only two-thirds the size of our own Moon. And, with the discovery of Pluto's largest satellite, Charon, in 1978, its mass became a measurable quantity – turning out to be far too small to have any effect on the orbits of Uranus and Neptune.

Of course, our knowledge of Pluto and its five moons has grown immensely thanks to the epic fly-by of NASA's *New Horizons* spacecraft on 14 July 2015. Far from being a dead world pock-marked with craters, as was expected, the dwarf planet is geologically active, with glacial flows of nitrogen slush, gigantic floating shards of water ice, and perhaps ice volcanoes. All this, despite an average surface temperature of around −233 °C.

The disappointment in Pluto's half-pint dimensions in the middle of the 20th century briefly spurred a renewed search for a hypothetical ninth planet that would be the scapegoat for the irregularities in the orbits of Uranus and Neptune. But by then, some astronomers doubted there was any need for such an object. Eventually, in the 1980s, the idea of 'Planet X' disappeared altogether, when the mass of Neptune was carefully measured from the trajectory of another famous interplanetary spacecraft, *Voyager 2*. That re-evaluation brought everything back into balance, and eliminated the need for any postulated new worlds. Surprise, surprise – the discovery of Pluto had been nothing more than a fortunate accident.

SO HERE WE ARE, EARLY IN THE 21ST CENTURY, FACED WITH a similar prediction of a large ninth planet out there in the furthest reaches of the Solar System. Should we believe it? My guess is yes. Why? First, the team that has published the prediction is led by Mike Brown of Caltech, one of the most prolific discoverers of objects in the Kuiper Belt and beyond. This man doesn't make predictions lightly, and his work also has the support of earlier proposals for the existence of a remote ninth planet. Second is the fact that this prediction is made on the basis of studies of the orbits of not one, but several very distant Solar System bodies. And that number is increasing. As a result of the intensive search for Planet Nine, other distant objects with suspicious orbital characteristics have been discovered. They sport charismatic names like 2012 VP_{113}, 2014 FE_{72} and 2015 TG_{387} – although the latter is also affectionately known as 'The Goblin', because it was discovered close to Halloween. And third, as I mentioned earlier, the hypothesised planet would not only solve the problem of the aligned eTNO orbits, but some other oddities in the Solar System as well – things like the slight tilt of the Sun's rotation axis relative to the orbits of the planets, and the occurrence of some eTNOs with almost 'vertical' orbits.

If Planet Nine is found, what do we call it? One suggestion, echoing Herschel's name for Uranus, is 'George'. But Mike Brown and Konstantin Batygin have also used 'Jehoshaphat', which, between themselves, they abbreviate to 'Phattie'. Some astronomers have objected to the term 'Planet Nine' itself, claiming that it is culturally insensitive since it diminishes the legacy of Clyde Tombaugh in discovering Pluto – which, at the time, was believed to be the ninth planet. They would prefer something with less implied bias, such as Planet X. In the end, the object will only need a formal name once it has been unequivocally identified. As always in such matters, that will be bestowed by the übernerds – the International Astronomical Union – and will almost certainly come from Greek or Roman mythology.

At the time of writing, a number of large telescopes are involved in the search for Planet Nine, and new, larger telescopes, soon to come online, will also take up the quest. It's entirely possible that by the time these words appear in print, Planet Nine will have been found, perhaps rendering much of this chapter redundant.

THE UNIVERSE
AT LARGE

CHAPTER 15
NATURE'S BARCODE: A USER'S GUIDE TO LIGHT

It's remarkable that astronomers can tell precisely what stars are made of, even though they cannot extract physical samples from them. The way they learned to do this is one of the great stories of astronomy, ranking in significance alongside the invention of the telescope. It starts with the unexpected villain of the piece, a 19th-century Parisian philosopher by the name of Auguste Comte. In many respects, he is a hero of science, since his philosophy of reason and the importance of rigorously testing ideas is at the heart of the scientific method. In 1835, however, he let himself down in regard to our understanding of the stars, when he confidently asserted that we would 'never be able by any means to study their chemical composition', and that such attributes as their density and temperature would be 'forever denied to us'.

Well, never say never. Particularly since, in that same year, scientists were already taking steps to understand the means by which we might investigate those matters. In August 1835, the English scientist Charles Wheatstone carried out a telling demonstration at the fifth meeting of the British Association for the Advancement of Science, held in Dublin. He used a prism, whose ability to deconstruct sunlight into a band of rainbow colours from deep violet to deep red had been established by Isaac Newton 170 years earlier, leading Newton to invent the term 'spectrum'.

153

But instead of using the prism to split sunlight into its component colours, Wheatstone pointed it towards an electric spark formed between two metal electrodes. Rather than a spectrum composed of a continuous band of colour, the prism revealed a set of discrete narrow lines of light, each an image of the spark itself, but composed of a single colour. It was as if the other colours between the lines had been rubbed out. We call these features 'emission lines', and now know that each one corresponds to light of a differing microscopic wavelength, with the violet lines having wavelengths about half those of the red. So, while plain white light, from, say, an incandescent lamp, is composed of gazillions of adjoining wavelengths producing what is predictably known as a continuous spectrum, the discrete bright lines of the emission spectrum are produced by the atoms of the metal excited by the spark.

Different metals emit different patterns of bright lines, as Wheatstone gleefully pointed out in Dublin. And that is the key to being able to determine remotely what stars – and many other classes of celestial object – are made of. In fact, earlier work by other English scientists had already shown that differing salts burned in a flame also produced differing sets of emission lines, but it was Wheatstone's demonstration that created interest in the topic.

Then, barely two years after Comte's death in 1857, a not-quite-household-name physicist at the University of Heidelberg by the name of Gustav Kirchhoff carried out a detailed analysis of the subject. He worked closely with a chemist who really is a household name – Robert Bunsen, of Bunsen burner fame. Together, they devised an improved device for viewing the spectra of light sources – a spectroscope – and used it to make the crucial discovery that every element has its own unique emission-line spectrum, not just a few metals. It is as if nature itself has hidden an identifying barcode in the light of every chemical imaginable. Once that barcode has been revealed by the spectroscope, the identity of the material is known.

There is a well-known photograph of these two great scientists standing together, probably taken early in their collaboration during the 1850s. The statuesque Bunsen towers over his younger, more slightly built colleague, giving them the vaguely comical appearance of a Laurel and Hardy of spectroscopy. Be that as it may, their collaboration yielded the fundamental rules of light analysis, embodied in what are known as Kirchhoff's Laws.

Briefly, they are (1) that incandescent bodies such as a white-hot lump of metal or a glowing electrical filament emit a continuous spectrum, (2) that materials excited in a spark or flame emit their own characteristic emission-line spectrum, as we have seen, and (3) that if you view the continuous spectrum of a hot object through a cooler gas, you'll get what is known as an absorption spectrum. What's that? Almost miraculously, the colours (that is, wavelengths) that would be emitted by the gas if it was excited are *subtracted* from the continuous spectrum of the background source, producing a ribbon of colour crossed not by bright lines, but dark ones. Not surprisingly, they are called absorption lines, since the light of the background source has been absorbed at those wavelengths – absorbed by the intervening gas. And, once again, the pattern of dark lines unambiguously identifies the gas through which the light has travelled.

In 1861, Kirchhoff and Bunsen were able to show that dark lines in the continuous spectrum of the Sun – recognised since 1802, but never understood – were absorption lines produced by known elements in the Sun's atmosphere. The atmosphere is at a lower temperature than the underlying luminous gas, which, by the way, is called the photosphere – the visible 'surface' of the Sun. At a stroke, the two scientists had definitively established what the

Sun is made of, despite the intervening 150 million kilometres. It is mostly hydrogen, but the spectral signatures of many other elements are there, too. The confident proclamations of Auguste Comte were, by now, seriously under threat.

THE FINAL BLOW CAME LATER IN THE 1860s. ANOTHER Englishman, a fellow subsequently aided by a wife who was at least as able as he was, had sold his family business in order to pursue his interest in astronomy. Equipped with a telescope capable of serious research, he greeted the news of Kirchhoff and Bunsen's work on the solar spectrum with enthusiasm, and resolved to investigate whether stars showed the same kinds of spectroscopic signatures as the Sun. His name was William Huggins, and he enlisted the help of a friend – a professor of chemistry from King's College, London, by the name of William Miller. Together, Huggins and Miller built a spectroscope for the telescope, and then embarked on a tour of the heavens, checking out everything bright enough to reveal a spectrum to their eager eyes.

What they found amazed them. While the Moon and planets exhibited essentially the spectrum of the Sun (as expected, given that they shine by reflected sunlight), the spectra of stars varied significantly. We now recognise that this is due principally to differing sizes and temperatures, a subtlety unknown to Huggins, but he had no difficulty grasping the main message. The barcode signatures of familiar earthly elements were there before his eyes in the absorption lines of the stars. As he later wrote, 'a common chemistry ... exists throughout the universe'. What a breakthrough. Huggins and Miller published their catalogue of the spectra of 50 stars in 1864, and the new science of astrophysics was born.

It used to be thought that Huggins' wife, Margaret, whom he married in 1875, only assisted him, but recent studies have shown clearly that they were an equal partnership, with several jointly authored papers to their credit. Moreover, Margaret's technical interests, which predated her marriage to William, enabled her to make innovations that significantly furthered their research. She was the one, for example, who promoted the idea of photography in studying the spectra of the stars, attaching a camera to a spectroscope to make what is still known as a spectrograph. Today's instruments are equipped with state-of-the-art electronic sensors rather than photographic plates, and are as sensitive as the laws of physics allow. But they work on the same principle as Margaret Huggins' spectrograph.

OF THE HUGGINSES' DISCOVERIES, WE SHALL HEAR MORE in this chapter, but there was one observation that eluded them. It had been known since the 1840s that starlight should exhibit something known as the Doppler effect. Most people are familiar with it, even if they might not be able to put a name to it. When applied to sound waves, it's the change in pitch that occurs when a sound source moves, most commonly heard when a fire truck or ambulance speeds by with its siren blaring. The sound is higher pitched when the emergency vehicle is approaching, and lower as it recedes, and the effect is caused by the wave-motion of sound.

The fact that exactly the same thing happens with light waves means astronomers can measure the speeds of objects along the line of sight, whether these are planets, stars, galaxies or whatever. They look for a shift in the spectrum lines, and by measuring it, can deduce the object's velocity in the *radial* direction (that is, towards or away from us – 'towards' producing a blue-shift, and 'away' a red-shift). The spectroscope or spectrograph in effect becomes a celestial speedometer.

These are delicate measurements to make, however, and while the Hugginses attempted them several times from 1868, it was not until 1889 that Hermann Carl Vogel, Director of the Astrophysical Observatory in Potsdam, obtained the first reliable measurements of stellar radial velocities photographically. In fact, Vogel's initial work concerned the bright star Algol in the northern-hemisphere constellation of Perseus. This star varies in brightness, and was already known to be what is called a binary system – that is, two stars orbiting a common centre of mass. Vogel detected a periodic shift in the spectrum lines, which he correctly interpreted as being due to the brighter of the stars exhibiting radial velocity changes as it orbited its companion. Such objects are known as spectroscopic binaries, because it is usually the case that the regular velocity changes are the only symptom of their duality. In their visual appearance, they are indistinguishable from single stars.

LET ME MENTION A FEW MORE APPLICATIONS OF THE spectroscopic technique. As Vogel's work on Algol suggests, the Doppler effect can be used to deduce whether – and how fast – things are rotating. We saw a few chapters ago that it was used in the 1890s to show that Saturn's rings rotate not like a solid object, but as a swarm of particles. The technique extends across the whole gamut of astronomy, from rotating planets, stars and gas clouds to whole galaxies of billions of stars.

Astronomers today are also using the effect to find things that are completely invisible. The planets of stars in the Sun's neighbourhood are, for the most part, too faint to see directly, even with the largest telescopes, but they can reveal themselves by the way they tug on their parent stars as they orbit around. The resulting backwards and forwards component of the star's motion is very small, ranging from several metres per second in the case of

Jupiter-sized planets to just a few centimetres per second for Earth-like objects. Despite that, the velocities can be detected with advanced equipment, and this so-called 'Doppler wobble technique' is routinely used at several of the world's major observatories, including the Anglo-Australian Telescope at Siding Spring. Actually, the biggest problem is calibration, since you have to compare super-precise observations taken days or sometimes weeks or months apart as the planets move around their parent stars, and you need to be sure that all the spectrum lines are measured against the same zero-point. Some novel and exotic optical devices are used for this – iodine cells and photonic combs, for example.

You might also be surprised to learn that magnetism can be detected by its effect on light. It was a Dutch physicist, Pieter Zeeman, who noticed in 1896 that spectrum lines (both emission and absorption lines) split into several components when the light is emitted in a magnetic field. This so-called Zeeman effect allows astronomers to probe the magnetism of the Sun and stars. And, by combining the Zeeman effect with the Doppler shift, it is possible to make maps of magnetised spots on stars (like the sunspots visible on the Sun), even though the stars are too far away for their discs to be visible. This complex but highly effective technique is called Zeeman Doppler imaging, and it's also carried out at the Anglo-Australian Telescope, principally by colleagues from the University of Southern Queensland.

And, finally, there's the expansion of the Universe. In 1929, American astronomer Edwin Hubble used a spectrograph to discover that galaxies are flying away from us with speeds that are proportional to their distances. Rather than being due to the Doppler effect (which is caused by the motion of an object *through* space), these so-called 'recession velocities' are interpreted as being due to the Universe itself expanding. In other words, space is getting bigger, and it carries the galaxies along with it. In homage to its discoverer, we call that overall expansion the Hubble flow.

Because the light from these galaxies has been travelling for hundreds of millions if not billions of years, the Universe has expanded significantly since it was emitted. The light waves themselves have participated in the expansion, so they arrive at our telescopes stretched to a longer wavelength than when they set out. That means the light spectrum – including the barcode of emission or absorption lines – is shifted to the red. This effect is called the 'cosmological redshift' to distinguish it from the simple Doppler effect, and it is one of the most remarkable tools available to astronomers.

Its effect is to date-stamp the light with the time when it was emitted. Because we know what the barcode of spectrum lines looked like when it left its source, we can measure directly how much redshift it has experienced. Thus, astronomers can deduce how much smaller the Universe was when the light set off, relative to today's Universe. And, knowing how the size of the Universe changes with time, they can calculate when the light left the galaxy that emitted it. Once again, this work is a major part of investigations at the Anglo-Australian Telescope, where the technique is used to make detailed three-dimensional maps of the Universe. They reveal the structure imprinted by the Big Bang – the event in which the Universe is believed to have been created some 13.8 billion years ago. And they are also being used to investigate some of the most pressing questions in modern astronomy, concerning the nature of dark matter and dark energy.

THE CONSTRUCTION OF SUCH DETAILED MAPS INVOLVES one further trick of the trade, and it's something I've been deeply involved with during my career in astronomy. When Huggins, Hubble and countless other astronomers throughout history made their spectroscopic observations of stars and galaxies, they had no alternative but to make them one at a time. And each observation took approximately forever. I have always admired the work of an early 20th-century American astronomer called Vesto Slipher, who carried out some of the observations of galaxy spectra used by Edwin Hubble in formulating his work on the expanding Universe. Slipher's catalogue of the spectra of 25 galaxies, published in 1917, required between 20 and 40 hours of observing for each galaxy. That meant observing the same object night after night to build up enough information on one of the crude photographic plates then in use, before developing it to reveal the faint spectrum.

Today's catalogues of galaxies are measured in hundreds of thousands, and will soon be in multi-millions. And the same is true of catalogues of stars in our own Galaxy. How are such totals achieved? The amount of exposure time per observation has fallen from tens of hours to tens of minutes by virtue of bigger telescopes, more efficient spectrographs and super-sensitive electronic image-sensors. However, even with such advances, astronomers would still be limited to observing their targets one at a time if it were not for the trick of the trade I spoke of. And that is to use clever technology to permit astronomers to observe hundreds of objects at a time – which, very soon, will increase to many thousands.

Most large telescopes have a reasonably wide field of view – that is, they see a significant chunk of sky with each observation. In a truly wide-field telescope like the United Kingdom Schmidt Telescope at Siding Spring Observatory, you can see an area of sky six degrees across – a dozen times the diameter of the full Moon, and big enough to encompass the whole Hyades Cluster. Most telescopes have a smaller field size than this, but you get the idea.

So – the telescope is really effective at presenting you with images of a large number of target stars or galaxies in its field of view, but how do you transfer individual samples of their light into the spectrograph? The answer is with optical fibres, thin strands of glass-like material that are as flexible as guitar strings, but transport the light from one end to the other with virtually no loss of intensity. If you have a bundle of hundreds of fibres, and can position one end of each accurately on a selected object, the flexibility of the fibres allows you to take the light to a convenient and stable location, often metres from the telescope, and then arrange the whole lot neatly in a straight line at the other end. Why? Because that is just what is needed to analyse them all simultaneously in a spectrograph. And the trick works like a dream.

The one difficulty with the multi-fibre spectroscopy technique is that each fibre has to be aligned exactly with its selected target in the telescope. That has to be accurate to a tiny fraction of a millimetre, and demands sophisticated robotic technology which has taken several decades to perfect through successive phases – most of which I have been directly involved with. Some of my colleagues have kindly referred to me as one of the pioneers of multi-fibre spectroscopy, and I guess it's true that I was the first to do various things, like using the technique to observe stars rather than galaxies (in 1982), and building some ground-breaking instruments for various large telescopes during the 1980s and 90s. I can't lay claim to writing the world's first PhD thesis on the topic, though. That honour goes to an eminent US astronomer by the name of John Hill. Mine was the second.

I don't think it's too immodest to say that the discoveries that have been made using this technique have revolutionised astronomy, with multi-fibre instruments now being used on most of the world's biggest telescopes. I've already mentioned large-scale galaxy surveys that help us understand both the way galaxies evolve and the way the Universe as a whole has evolved. But the spectroscopic observation of large numbers of stars is giving us similar insights into the structure and evolution of our own Milky Way Galaxy. From 2003 to 2013, the UK Schmidt Telescope I mentioned earlier was occupied by a survey called RAVE – the RAdial Velocity Experiment. Remember radial velocities? I was the project manager for RAVE, and I'm delighted that the final catalogue of half a million star velocities and other characteristics is just about to be published. Meanwhile the larger Anglo-Australian Telescope is undertaking a survey of a million stars known as GALAH – which might sound like a much-maligned Australian parrot, but is actually GALactic Archaeology with HERMES. Of course, HERMES is itself an acronym for the super-sensitive home-grown spectrograph being used, while galactic archaeology is the investigation of the history of our Galaxy by measuring the exact chemistry of as many stars as possible. And there are new surveys in the offing, using new technology that will extend the capabilities of multi-fibre spectroscopy well into the 2030s. I feel privileged to have been so closely involved with this revolution.

LET ME RETURN FINALLY TO ONE OTHER DISCOVERY MADE by William Huggins. While it wasn't really his fault, it became an example of unjustified scientific hubris rivalling that of Auguste Comte. In the early 1860s, when Huggins was wondering whether there was anything his new science of astronomical spectroscopy couldn't achieve, he turned his attention to one of the great scientific problems of the time. That concerned the nature of nebulae – ill-defined misty patches in the sky that were neither stars nor planets. The Big Question was whether they were made of myriads of stars too faint to be seen individually, or something else, such as a cloud of glowing gas. Or (as we now know to be the case), an assortment of both.

Huggins directed his spectroscope towards one of these nebulae in the August of 1864, and was amazed by what he saw. Emission lines – the clear signal of an excited gas – rather than the absorption-line spectrum of a cloud of stars. As he recalled three decades later, 'The riddle of the nebulae was solved. The answer, which had come to us in the light itself, read: Not an aggregation of stars, but a luminous gas.' At the age of 40, Huggins had revolutionised the astronomy of his time, and his place in history was assured.

But there was a snag. Very soon, astronomers realised that some of the emission lines they could see in various nebulae didn't belong to any known element on Earth. Yes, hydrogen was there, but what was this bright green line that didn't correspond to anything they'd seen already? And others? It was, indeed, a puzzle – but there was a precedent in a strange yellow line that had been observed in the spectrum of the Sun during an eclipse in 1868. Two English scientists, Norman Lockyer and Edward Frankland, had deduced that this was the signature of an unknown element that was present in the Sun, but not on Earth. Dubbed 'helium', it was expected to reveal itself some day in the inventory of terrestrial chemical elements. And so it did – in 1895, in the hands of a Scottish chemist by the name of William Ramsay, who isolated it from a mineral known as cleveite. It was the first chemical element to be discovered in space rather than on Earth – a triumph for astronomical spectroscopy.

It's no wonder, then, that astronomers should take it for granted that the unidentified emission lines in the spectrum of nebulae, including the mysterious green line, were the spectral signature of another unknown element. With supreme confidence, they dubbed it 'nebulium', a name Margaret Huggins first recorded in 1898, but probably did not invent. Using the measured wavelengths of the lines, and improvements in the understanding of atoms, scientists worked hard to discover the properties of nebulium. In an impressive research paper published in 1914, for example, a trio of eminent French astronomers even deduced that it must be two different elements, but got no nearer to identifying what they were.

At last, with improvements during the early 20th century in our understanding of why emission lines occur at all, the mist started to clear. The fact that excited atoms emit light with very specific wavelengths comes about because of specific energy levels occupied by the electrons clouding around their nuclei. Particular wavelengths are emitted when the electrons jump from one energy level to another, emitting a certain 'quantum' of light. Sound familiar? Yes, it's the foundation of quantum theory. But one of the theory's quirks is that it incorporates so-called selection rules. Some of those energy transitions are permitted, while others are forbidden. They don't happen. Actually, they do, but only if the excited atoms belong to a gas at a pressure much, much lower than anything possible in a laboratory on Earth. A rarefied gas in the depths of space, for example.

It was a 28-year-old genius at the California Institute of Technology by the name of Ira Sprague Bowen who, in 1927, was busy calculating the theoretical wavelengths of light that would be emitted by the electron transitions of various elements. Of course, he followed the selection rules – until he realised that the forbidden lines weren't really forbidden, but just extremely unlikely at the gas pressures encountered on Earth. In a moment of brilliance, he thought of nebulium. Going back to his calculations, he worked out what forbidden emission lines might be emitted if the selection rules didn't prohibit them. And sure enough, when he looked at oxygen, the forbidden wavelengths matched those of nebulium perfectly – including that bright green line. Eureka! Inspired, Bowen feverishly calculated the forbidden lines

THE UNIVERSE AT LARGE

that other elements would emit, and obtained similar outcomes. His results eventually appeared in a seminal paper in 1928. And nebulium was consigned to the history books. Ira Bowen went on to have an outstanding career in the astronomy of the United States, masterminding some huge advances in both the science and technology of astronomy, until his death in 1973.

CHAPTER 16

REVERBERATIONS:
EXPLODING STARS
AND LIGHT ECHOES

Our Moon travels about 88 000 kilometres through space in a day, as it orbits Earth. That's way more than most of us drive in a year. Admittedly, most of us aren't averaging 3679 kilometres per hour, like the Moon is, but it's still a heck of a long way. And Earth is moving in its orbit around the Sun – taking the Moon along with it, like the dutiful parent it is. In one day, Earth hurtles through 2.6 million kilometres, more than 200 times its own diameter. But then the Sun and its planets are moving holus-bolus around the centre of our Milky Way Galaxy as part of a gigantic swirl of stars embellished with graceful spiral arms. Between noon on one day and noon on the next, the Sun and its entire retinue cover an impressive 20 million kilometres. Which just goes to show that space, as Douglas Adams so eloquently pointed out, is big.

But, accepting that space is big and things move a long way through it, what else in the Universe happens on a time-scale of one day? Curiously, the answer is not much, apart from the occasional cataclysmic event. A pair of black holes merging, for example, or an asteroid hitting the surface of a planet. Such things certainly produce rapid changes in their immediate surroundings, but generally speaking, the Universe is pretty much the same from one day to the next. Most of what happens takes place on time-scales comparable with Earth's geological processes – timescales of tens or even hundreds of millions of years. Thinking about them means resetting your mental clock into a different regime.

PARADOXICALLY, THOUGH, IT'S SOMETIMES THE BRIEF cataclysmic events that initiate long-term changes in the Universe. Take exploding stars, for example – the objects we know as supernovae. When a star more than about eight times the mass of the Sun reaches the end of its life, it detonates with extraordinary energy. The explosion initiates in a split second, but the processes involved take a few hours to develop. Even though the supernova's rise to its ultimate brilliance then takes a matter of days or weeks before it starts to fade back into oblivion, this is the blink of an eye compared with most other events in the Universe. In fact, the way a supernova evolves and the amount of energy it releases depend on the exact type of explosion it is. Astronomers now recognise a score of different varieties, involving a range of masses and triggering mechanisms.

One of the consequences of a supernova – which is of paramount importance to our own existence – is the creation of new chemical elements in the Universe. You might be aware that the carbon, neon, oxygen and silicon in our world were forged in the interiors of ordinary stars like the Sun. Those elements came from nuclear processes that start with hydrogen, which was created in the mother of all cataclysmic events – the Big Bang, some 13.8 billion years ago.

But many elements more massive than iron can only have been engineered in the extremes of temperature and pressure encountered in a supernova explosion – an idea first advanced back in 1954 by the same Fred Hoyle we met in connection with panspermia. In fact, it was this discovery that set Fred on the road to astronomical stardom. A gruff Yorkshireman who I knew personally only late in his life, he set the tone of postwar astronomy, despite espousing a theory of the Universe that eventually turned out to be wrong. He was a lifelong champion of the 'steady state' theory, which maintained that matter is continuously being created as space expands, in contrast to the alternative idea that the Universe was created in a single massive explosive event. Fred mocked this as the 'Big Bang' – and yes, that's where the name came from.

Fred Hoyle's inventory of atoms created in supernova explosions include trace elements essential to our own health, as well as some beguiling elements that we prize because of their rarity. Gold and platinum, for example – not to mention a few slightly less beguiling items like uranium and lead. But it's surely a fascinating thought that some of the contents of your jewellery box started their existence in the unbridled fury of an exploding star.

Another key consequence of these explosions is the shock waves they send ringing through their surroundings. The blast of material emitted by the supernova sweeps up the rarefied gas between the stars – the so-called interstellar medium, whose density is normally so low that it's measured in individual atoms per cubic centimetre. But once compressed, it can spawn a new generation of stars – particularly if the shock wave passes through one of the denser clouds of gas and dust that abound in the disc of a galaxy like our Milky Way. The more massive members of that new generation are blue-white stars that are extremely bright, and totally profligate with their reservoirs of hydrogen fuel, so that their brief lifetimes are counted in millions rather than billions of years. Live fast, die young. Thus, they themselves quickly turn into supernovae, producing another shock front that can then progress through the disc of the galaxy.

It's this mechanism, carried to extremes, that gives rise to the beautiful spiral arms we see in many galaxies. Surprisingly, they are nothing more than a grand illusion. They are traced out, not by strings of run-of-the-mill stars – which are actually rather evenly spread throughout the disc of the galaxy – but by strings of moderately rare, but intensely bright young stars, whose existence has been triggered by the supernova-driven wave passing through the galaxy's disc. This curious groundswell of star formation is known as a density wave, and is effectively a sound wave passing through the rarefied material of a galaxy – a sound wave that is revealed by the young stars it has given birth to.

SUPERNOVAE CAN PRODUCE ANOTHER FAMOUS ILLUSION, and this is one that has fascinated me since I first encountered it 30 years ago. Let me introduce it by asking you to imagine an echo. What do you think of? A shouted 'Coo-ee', reflected from a distant cliff face, perhaps? Or the dying reverberation of music in a great cathedral? Most of us love those bouncing sound waves, and the sense of ambience they create. But you might be surprised to learn that astronomers are very fond of another type of echo – one that involves not sound, but light. And, remarkably, we can use it to map out the structure of dusty regions in our Galaxy. Or look deep into the past to see long-dead supernova explosions as if they were happening today.

When early astronomers explored the night sky with telescopes, they found among the stars and planets multitudes of small misty patches that they called 'nebulae' – from the Latin word for fog. It was a marvellously generic term for something whose true nature was unknown. We now recognise several different kinds of nebulae, but one particular type is made up of smoke-like particles of dust that reflect the light from nearby stars. Not surprisingly, astronomers call them 'reflection nebulae'.

Now imagine a reflection nebula lit not by the constant glow of starlight, but by the searing flash of a supernova. The nebula might have been invisible beforehand if there were no nearby stars to illuminate it, but it will brighten into visibility as the supernova's light reaches it. Because light travels at a finite speed through space, and because space is so big, it could light up months or years – or even centuries – after the explosion occurred. This extraordinary effect is called a light echo. And since the supernova's pulse of light may last for only a few weeks or months before it fades back into obscurity, the light echo is exactly analogous to the audible echo of a short burst of sound. Just like your cheerful 'Coo-ee' bouncing off the distant cliff.

One other important attribute of a supernova's light pulse is that it radiates in all directions, surrounding the exploding star with an expanding shell of light. That means it can bounce off any nearby dust clouds, no matter where they lie in relation to the supernova – in front of it, behind it, or off to the side. The key point is that clouds at different distances from the supernova will usually light up at different times. I say 'usually', because there is a subtlety here concerning the time delay of the light echo compared with the direct pulse of light travelling in a straight line to Earth. That time delay depends not only on the distance of the reflection nebula from the supernova, but also its distance from us. Any two dust clouds for which the sum of these distances is equal will therefore light up simultaneously – because they have the same time delay, as the light travels along its dog-leg path.

With all that information, it's possible to accurately map the distribution of dust in the vicinity of a supernova, as the light pulse bounces off each individual cloud, or each part of a cloud. It's rather like the booming reverberation of a nearby thunderclap from buildings or the ground, except that in the supernova light echo, the source is a single intense point of light rather than the

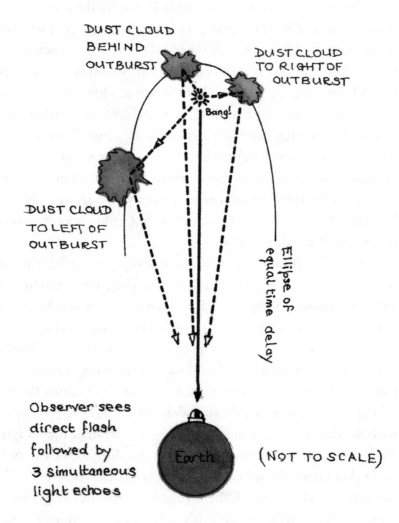

Light from an outburst such as a supernova explosion can reach our telescopes either directly (solid line) or via reflections from dust clouds (broken lines). These light echoes arrive later, but any dust cloud lying along the elliptical line will be seen with the same delay. In three-dimensional space, the shape is an ellipsoid.

Author

tortuous line of superheated plasma created by a lightning bolt. Which means that the geometry of the dust cloud can be exactly calculated from the light echo, providing astronomers with a powerful investigative tool.

PERHAPS THE BEST-KNOWN EXAMPLE OF A SUPERNOVA light echo was the one resulting from Supernova 1987a, which was seen to explode early in 1987 in the Large Magellanic Cloud, or LMC. The LMC is one of our nearest neighbour galaxies, visible from the southern hemisphere as a large fuzzy patch (definitely not a small misty one), looking to the unaided eye just like a broken-off bit of the Milky Way. For a few weeks, the supernova was clearly visible without a telescope, becoming the first naked-eye supernova since the one seen by the great German mathematician and astronomer, Johannes Kepler, late in 1604. (The tangled remains of Kepler's supernova are still visible, by the way, in the constellation of Ophiuchus. We grace such exotic debris with the technical name of 'supernova remnant'.)

Of course, Supernova 1987a didn't actually explode in 1987. The event had happened some 160 000 years beforehand. Such is the distance of the LMC that the light pulse travelling directly towards us from the supernova at 300 000 kilometres per second took that long to get here. Naturally, it's why we describe its distance as 160 000 light years.

The arrival of the light pulse took the world's astronomers by surprise – as you might expect, given that supernova explosions are impossible to predict. One of them – my famous colleague, Rob McNaught, of Siding Spring Observatory – was more devastated than surprised. He had photographed the LMC as part of a routine survey on the night the supernova appeared, but exhaustion at the end of the night had led him to delay processing the film until the next morning. I remember him turning up in the Schmidt Telescope building the following afternoon, looking slightly dazed.

'There's a naked-eye supernova in the LMC,' he said bleakly. 'The first naked-eye supernova for nearly four hundred years. And I missed the discovery.' He was undaunted in his pursuit, however, and was subsequently named as 'the world's best observer' by a prominent US astronomy magazine.

Supernova 1987a sent the world's astronomers into a frenzy of activity. For the first time, a supernova bright enough to be seen with the unaided eye could be scrutinised with the battery of sensitive optical instruments at the modern astronomer's disposal. The Anglo-Australian Telescope was in pole position, and special instrumentation was hastily knocked up by its resourceful staff. In astronomy, as in other sciences, it pays to be nimble on your feet, and this was a classic example. As a result, the supernova was measured, fathomed and analysed to within a whisker of its life, teaching us more about supernovae than we had learned in the previous 50 years.

But then, a couple of years after it had faded from brilliance, something odd happened. Just as astronomers were breathing a sigh of relief that the excitement was over, and they could return to their normal studies, a specially processed photograph made by astrophotographer David Malin with the Anglo-Australian Telescope revealed the presence of two faint rings around the supernova remnant. They excited the interest of a colleague of ours, an astronomer by the name of David Allen, who was one of the most energetic and gifted scientists of his day. David was a true polymath of astronomy, with research interests ranging from the atmosphere of Venus to the most distant objects in the Universe – and a knack of being able to talk engagingly about his discoveries to scientists and non-scientists alike. It's no accident that the

Astronomical Society of Australia's award for communicating astronomy to the public is named in his honour.

David realised immediately that the rings that had appeared around the supernova were not caused by bubbles of expanding material. At the distance of the Large Magellanic Cloud, such bubbles would need to have swelled faster than the speed of light to attain their observed diameter – which, is, of course, impossible. But accurate photographic measurements of the rings made by the two Davids and their colleagues soon proved that they were echoes of the supernova's light. Their colour exactly matched that of the supernova at its brightest.

Why did this light echo take the form of rings of light, rather than the individual blobs that would appear as the supernova illuminated nearby dust clouds? Once again, David Allen was quick to realise what was happening. The supernova was illuminating not individual clouds of dust around the supernova, but something semi-transparent lying in front of it. David wrote a vivid account of his research in a 1991 British publication called the *Yearbook of Astronomy*. His article describes in detail how he showed that the rings are caused by light scattered towards us by two thin sheets of dust in front of the supernova, a scenario that would be far from obvious to the casual observer. You can sense the excitement he felt as he made the calculations that revealed this geometry, and his satisfaction when he realised that the sheets are probably the front and back surfaces of – yes, a dust bubble in space. But this was an enormous bubble, lying a very long way in front of the supernova, and completely unrelated to it.

David's hypothesis was that a cluster of hot stars with the slightly inelegant name of NGC 2044 had excavated this bubble by outflowing material from the individual stars piling up the surrounding matter ahead of it 'like a snow plough, building up an ever denser ripple as it goes'. Since this collection of hot stars lies in front of and slightly to one side of Supernova 1987a, that seems perfectly reasonable. We now know from modern images that the entire region is veined with gas and dust clouds, and that gigantic bubbles of dusty material from ancient supernova explosions are everywhere.

David also predicted that Supernova 1987a would excavate its own bubble in space, but one that would be elongated, rather than spherical – a so-called bipolar nebula. And he was right on the money. Only a few years after his *Yearbook of Astronomy* article was published, images from the then-new Hubble Space Telescope showed a bright ring of light, where the ejected material from the supernova was beginning to collide with material shed more gently by the original star before it exploded. That collision is still ongoing, but we now know much more. Thanks to detailed observations made a few years ago with the European Southern Observatory's Very Large Telescope in Chile, the ring of mate-rial can be seen in three dimensions as the 'waistline' of an hour-glass-shaped bubble of debris. A bipolar nebula – exactly as David had surmised.

Sadly, David Allen did not live to see these exciting discover-ies. He died on 26 July 1994 at the age of 47, from a brain tumour. But I can well imagine the glee with which he would have wel-comed our present-day knowledge of Supernova 1987a.

I THINK DAVID WOULD HAVE BEEN THRILLED, TOO, WITH more recent observations of light echoes. In the early 2000s, a dust cloud was seen that was much more complex than the thin dust sheets in front of Supernova 1987a. The story began in January 2002, when a previously unnoticed star brightened to 600 000 times the Sun's luminosity before fading again. Because it was initially thought to be a fairly ordinary variable star – one that varies in brightness – it was given the standard and rather unmemorable designation of V838 Monocerotis. That means it was the 838th variable star discovered in the constellation of Monoceros, the Unicorn.

The outburst didn't constitute a supernova explosion – which would have blown the star to pieces – so much as an unprecedented and as-yet unexplained increase in its size. At its distance of 20 000 light years, V838 Monocerotis was unlikely to give up its secrets to anything less keen-sighted than the Hubble Space Telescope (HST). And, since 2002, repeated observations of the star with the HST have revealed that it is surrounded by a large and very complex dust cloud. As the sphere of light from the star's outburst has expanded, different parts of the cloud have been illuminated in a bullseye pattern of light and shade, with a filigree structure that is probably related to tangled magnetic fields. Once again, because the geometry of the light echo is well understood, astronomers can use it to probe the complex make-up of the dust cloud in a technique very similar to the computer tomography we're familiar with in the medical world.

During May to December 2002, the V838 Monocerotis nebula appeared to expand from four to seven light years in diameter, a so-called 'super-luminal' (or faster-than-light) expansion. As in the case of the Supernova 1987a light echo, this is an illusion, and results from the way the expanding light shell appears to illuminate its surroundings. More recently, NASA and the European Space Agency (ESA), which together operate the HST, have issued a remarkable video of the evolving light echo, made from morphing successive images of V838 Monocerotis obtained between 2002 and 2006. The video is well worth hunting out on the internet – but remember that although it looks like an expanding shell of material, this is just an impression.

In general, light echoes provide a kind of time machine that allows us to look back at outburst events similar to that of V838

Monocerotis, but which took place well before the modern era of astronomy. Recently, for example, astronomers have studied a long-gone outburst from a southern hemisphere star that is one of the most unstable objects in our Galaxy. Its name is Eta Carinae, and back in the 1840s, it was for a short while the second brightest star in the sky, with an intrinsic brightness some six million times that of the Sun. Good thing it's at least 6000 light years away. That outburst found its way into the Dreamtime legends of the Indigenous Boorong people of Victoria. But observations made in late 2014 and early 2015 have identified its light echo from a nearby dust cloud. The star was already known to be binary in form — two component stars orbiting around their common centre of mass. And the new observations suggest that the smaller component was actually immersed in the bloated outer atmosphere of its more massive companion during the outburst. No wonder things brightened up so much.

Even more intriguingly, light echoes can be used to study objects that shone brightly in the very distant past, but have now faded into obscurity. In 2008, astronomers using the Japanese Subaru Telescope in Hawaii observed light from dust clouds illuminated by a supernova that had lit up brilliantly for a few months in the early 1570s, when it was observed by the great Danish astronomer Tycho Brahe. While Tycho saw the direct pulse of light from the supernova in November 1572, the faint echoes were reflected from dust clouds a long way from it, adding an extra path length of 436 light years to its distance of around 9000 light years. I find it an extraordinary thought that modern high-tech analysis can now be applied to exactly the same brief outburst of light that Tycho had observed over four centuries earlier, revealing details of the supernova that he could never have imagined.

And more is to come. With a new generation of 'extremely large telescopes' just around the corner, we are sure to discover more of these distant light echoes from long-extinct supernovae. Perhaps even the spectacular daylight supernova recorded by Chinese astronomers in 1054 might yield its secrets to modern telescopes. The remnant of that explosion is the Crab Nebula – one of the best-studied objects in the entire sky. If we could examine the light of the explosion itself in similar detail almost a thousand years after it happened, it would be a remarkable achievement. As reverberations go, that's a pretty long one.

CHAPTER 17
SIGNALS FROM THE UNKNOWN: THE FAST RADIO BURST MYSTERY

In 2007 West Virginia University professor Duncan Lorimer and his colleagues were trawling through archived data from the Parkes radio dish in Australia, looking for the characteristic signals emitted by objects known as pulsars: brief, clock-like bursts of radiation that are extremely regular. These signals come from spinning neutron stars – incredibly dense and highly magnetised objects whose radio radiation sweeps through space like a lighthouse beam. As you might expect, if Earth is in the right direction to intercept the beam, the radiation arrives as a series of pulses.

What Lorimer's team found, though, was something different. In data that had been collected six years earlier, on 21 July 2001, there was a single, very strong and very brief pulse of radio emission. By brief, I mean less than five-thousandths of a second (5 milliseconds). A further 90 hours of data from the same part of the sky revealed no further bursts, however, suggesting that whatever caused the so-called 'Lorimer Burst' was a unique event – perhaps an exploding star, or a merging of neutron stars.

The team gleaned one other important piece of evidence from that observation. Just as white light consists of many rainbow colours mixed together, so radio radiation is composed of many different frequencies. And just as white light can be broken into its component colours by passing it through a prism, so radio radiation can also be spread into a spectrum of frequencies by clever receivers that observe them all simultaneously. It's like being able to sweep the tuning knob on an AM or FM radio instantaneously up and down to pick up all stations at the same time. When the team analysed radio frequencies in the Lorimer Burst they saw that they were smeared out slightly in time during the 5-millisecond burst, so that the higher frequencies arrived before the lower.

This 'dispersion' of frequencies is familiar to radio astronomers, who interpret it as being due to the passage of the radio signal through deep space that is not totally empty, but contains clouds of electrons that slow down the lower frequencies. And the amount of dispersion observed in the Lorimer Burst suggested that its source was at a distance measured in billions of light years – well beyond the environs of our Milky Way Galaxy. Whatever it was, the Lorimer Burst was a long way off – which implied that it was something very radio-bright indeed.

THEN, NOT LONG AFTER THE LORIMER BURST WAS DIS-covered, the story lurched unexpectedly into slapstick. It started with the realisation that comparable signals had been noted at the Parkes Radio Telescope since 1998. They had very similar dispersion characteristics to the Lorimer Burst – but these events seemed to be linked somehow to the time of day rather than being randomly distributed, as would be expected of signals from the distant Universe. With an alleged earthly origin, suggestions as to their possible source and how they might relate to the Lorimer Burst began pouring out. Lightning flashes, nuclear bomb tests and even aircraft radio signals were suggested. The events were given the name 'perytons', after a mythical creature invented by the Argentinian novelist Jorge Luis Borges.

But by 2015, phenomena more akin to the original Lorimer Burst had been observed at other radio telescopes, and they were clearly of celestial origin. Astronomers were starting to refer to them as Fast Radio Bursts or FRBs. Perytons, on the other hand, despite the elegance of their name, were becoming a tad suspicious. They had been observed only at the Parkes Radio Telescope, and nowhere else. Moreover, they seemed to be observed most frequently around lunchtime. Eventually, the penny dropped, and scientists realised that if the microwave oven in the Parkes lunch room was clicked open just before it had finished cooking, it emitted a burst of radiation that the nearby telescope could detect. Moreover, the sudden shut-down of the microwave's electronics produced a frequency dispersion that mimicked an FRB perfectly. At the cost of a few red faces among my radio astronomy colleagues, the peryton problem had been solved.

ONCE THAT RED HERRING WAS OUT OF THE WAY, THE hunt resumed in earnest for true FRBs, and again, several were discovered in archival data – this time from a range of large radio telescopes throughout the world. They were found all over the sky, supporting the notion that they are extremely distant, and bear no relation to our Milky Way Galaxy with its flattened disc of stars. Each detection of an FRB happened only once, suggesting that they all originated in a destructive event such as a pair of neutron stars colliding. Then, in November 2015, archival data gathered by what was then the world's largest single-dish radio telescope, at Arecibo, in Puerto Rico, revealed that one FRB had flared several times in irregularly spaced bursts of radiation.

This radio source, known by the splendid name of FRB

121102, clearly could not be bursting due to a destructive event. End-of-life explosions and hyper-collisions were ruled out by such a spirited performance. And FRB 121102's multiple burst activity has continued, with recorded events now numbering well over 100. Compounding the mystery is the fact that the multiple bursts have allowed astronomers to pinpoint its exact direction in space. As expected (since FRBs are assumed to be the result of something to do with stars – albeit exotic ones), the direction coincided with that of a galaxy whose distance can be measured. It's a cool three billion light years away. This means that in each burst, the object – whatever it is – pushes out more energy than the Sun radiates in a year. And then it does it again ... and again ... and again.

Realising that the multiple bursts of FRB 121102 mark it out as something different from other FRBs, astronomers shrugged their shoulders and concentrated their efforts on trying to understand the astrophysical mechanism of the common single-shot variety. Flares on highly magnetised neutron stars called magnetars have been suggested for their origin, as have collapsing pulsars and exploding black holes. Magnetism is a common theme in the various theories, but the bottom line is that FRBs remain one of contemporary astronomy's biggest mysteries. Based on observations so far, some astronomers have suggested that one occurs somewhere in the Universe every second, throwing our lack of understanding of these phenomena into stark relief.

On the positive side, the new technology becoming available will soon bring fresh information to bear on the problem. For example, in its first year of full operation, the Australian Square Kilometre Array Pathfinder (ASKAP) telescope array virtually doubled the number of known FRBs (ignoring the pathological repeater, FRB 121102). ASKAP is very well suited to the problem of finding them, since each of its 36 antennas is equipped with what amounts to a wide-angle radio image sensor known to its fans as a PAF (phased-array feed). While, in addition, the antennas can all be pointed in slightly different directions in so-called 'fly's-eye' mode. That allows ASKAP to image an area almost 1000 times that of the full Moon in its quest to find the elusive flashes. In this kind of work, field of view is as important as sensitivity, and the facility is now regarded as among the best in the world for FRB detection.

Among ASKAP's 19 newly discovered FRBs was one with the lowest dispersion yet found, meaning it was the closest that had been observed. You won't be surprised to hear that it rejoices in the name of FRB 171020. The dispersion measurement suggested that its distance was less than a billion light years – not on what you'd call our cosmic doorstep, but certainly within the range of visible-light telescopes if its host galaxy could be identified. And, by dint of a ten-year strategic partnership between Australia and the European Southern Observatory (ESO), signed in 2017, that's exactly what happened. At Cerro Paranal in Chile, ESO operates four 8.2-metre telescopes that are collectively known, not by some elegant European name, but as the VLT – the Very Large Telescope. They are equipped with the most comprehensive suite of instruments in the world, and are now accessible to Australian astronomers by open competition.

It was with one of them that a galaxy was found whose position matched that of FRB 171020 – a galaxy with another gobbledygook name: ESO 601-G036. We just can't help ourselves. Anyway, its distance, as measured by the VLT, is 120 million light years, well within the upper limit of the dispersion measure of the FRB. So, for the first time, the host galaxy of a non-repeating FRB had been identified. Why is this important? Because if we can do it for lots of FRBs, we may discover some property of their host galaxies that causes them. And ESO 601-G036 gives just a hint of this, in that there is another, very dim galaxy close by, one that might recently have been in collision with it. Galaxy collisions are violent events that stir up the gas in them, potentially forming brilliant and short-lived stars. Perhaps that might be found to be a common theme, when we know more about the galactic environments of other FRBs.

One of the astronomers leading this work is a friend and colleague from the former Australian Astronomical Observatory, Stuart Ryder. He is particularly excited by the success. 'We're standing on the cusp of an exciting new era,' he says, 'in which we are about to learn where Fast Radio Bursts take place. It is so fortuitous that this coincides with the start of Australia's access to ESO, bringing together the best radio and optical telescopes on the planet, in the best observing sites on two continents.' Fortuitous indeed, and the latest news from Stuart is that as of June 2019, the research group he belongs to has discovered another three FRBs, one of which, again, has an identified visible-light counterpart. This time, however the precision is high enough to identify in which part of the galaxy the FRB is located. It is well away from its dense central region, which has surprised some astronomers, since it suggests the FRB has nothing to do with the galaxy's central black hole.

Meanwhile, as if in echo of Stuart's words about new observational resources, a novel northern hemisphere radio telescope has also reaped a significant harvest of FRBs, including at least one further repeater – and possibly as many as five. CHIME is the Canadian Hydrogen Intensity Mapping Experiment, situated in British Columbia, and is sensitive to a lower radio frequency than ASKAP. What it boasts in sensitivity, however, it lacks in accuracy of directional location, so no visible-light counterparts have yet been identified. But colleagues at another new facility at Caltech's Owens Valley Radio Observatory announced another identification of an FRB with a distant galaxy in July 2019. These discoveries highlight the importance of new facilities in this rapidly developing field, and most of the world's astronomers can't wait for them to start shedding more light on the origin of Fast Radio Bursts.

THERE'S ONE ASTRONOMER WHO HASN'T WAITED, HOWever. And the frustration he has felt at the lack of a clear model for FRBs has led him to a particularly radical conclusion. This man is Avi Loeb of the Harvard-Smithsonian Center for Astrophysics, an eminent scientist not known for holding back on ideas that many astronomers consider off-limits. Loeb notes that 'because we haven't identified a possible natural source with any confidence, an artificial origin is worth contemplating and checking'.

In March 2017, Loeb and his co-author Manasvi Lingam of Harvard University speculated that FRBs might be caused by radiation from lasers being fired by extraterrestrial civilisations to drive light-sail-powered spacecraft through their own galaxies. They note that 'the beams used for powering large light-sails could yield parameters that are consistent with FRBs' – in other words, the physics holds up. Indeed, the physics holds up to such an extent that the work was published in no less a place than the *Astrophysical Journal Letters*, an academic publication that would have no truck with frivolous contributions. Moreover, Lingam and Loeb note that the multiple bursts of FRB 121102 could also be explained by their hypothesis. As, presumably, can the more recently reported repeaters.

Does Loeb really believe FRBs are the result of alien intelligence? 'Science isn't a matter of belief, it's a matter of evidence,' he says. 'Deciding what's likely ahead of time limits the possibilities. It's worth putting ideas out there and letting the data be the judge.'

It has to be said that most members of the world's astronomical community are sceptical. Maybe even scornful. But they might also like to note that a recently funded proposal to investigate whether microscopic spacecraft could be sent from Earth to the nearest star known to host its own planets – a project known as Breakthrough Starshot – relies on laser-driven light sails to power the spacecraft.

Just imagine how that technology might evolve over a few centuries – you could have FRBs all over the place.

CHAPTER 18
EYE OF THE STORM: BLACK HOLES INSIDE AND OUT

On 10 April 2019, a remarkable image grabbed the attention of the world's media. It showed the shadow of a black hole containing 6.5 billion times the mass of the Sun at the very heart of a distant galaxy, clearly defined by a telescope the size of Earth. For the first time, the predicted ring of radiation narrowly escaping the clutches of a black hole was visible at a magnification equivalent to reading newsprint from the opposite side of a continent.

Known as Messier 87, the target galaxy is some 55 million light years from our own Milky Way. It's known as an active galaxy, meaning that its central black hole is consuming gas and stars from its surroundings. But at present, it is relatively quiescent, allowing us to see the black hole's shadow.

The successful observations were made in 2017 using the 'Event Horizon Telescope', an array of eight high-frequency radio observatories spread around Earth's western hemisphere. Each was equipped with special data recorders, atomic clocks and sensitive detectors. For the experiment to work, the weather had to be good at all the sites. But in the event, out of a ten-day allocation of telescope time, astronomers required only seven days. The result was five petabytes of data – the equivalent of 5000 years' worth of MP3 plays – which have now been reduced to an image of a few kilobytes.

The feat involved a decade of work by a major international collaboration. At the media conference, project director Shep Doeleman paid tribute to the many scientists involved, with special praise for the early-career researchers who had carried out much of the work involved with the data reduction. Typical was the role of American computer scientist Katie Bouman, who had led the group developing a key algorithm in the imaging process. Her excitement at the result was captured in a gleeful image that quickly went viral on social media, prompting Bouman to counter the attention by emphasising how much the entire collaboration had contributed to the project

Asked whether there was a party once the final image had emerged, Doeleman admitted that the overwhelming emotion was surprise that the image was as expected. And the director of the National Science Foundation, France A Córdova, who had not seen the image before the media conference, confessed that it brought tears to her eyes. Such is the emotion generated by epoch-making science – which this undoubtedly was.

WHILE THIS HIGH-PROFILE MEDIA ANNOUNCEMENT succeeded in capturing global attention, much less evident to the public is the depth of knowledge that has been amassed over many decades in the study of black holes. Everyone loves them – from kindergarten kids to professors of theoretical physics – but until that image was released, much of our knowledge was founded on mathematical theorems developed to describe their expected properties. Now we've had first-hand evidence that those theorems work.

One of my favourites among them is something with the curious name of the 'no hair' theorem. It says that from the outside, the only observable properties of a black hole are its mass, its electrical charge and its angular momentum, or spin energy. All its other characteristics are hidden behind the veil of the event horizon – the boundary beyond which no radiation can escape. In other words, the black bit. The term was coined around 1970 by the American theoretical physicist, John Wheeler, who commented that 'black holes have no hair', meaning that all information other than that mentioned above is inaccessible to outside observers. Wheeler actually attributed the term to his student, Jacob Bekenstein, who worked with him at Princeton University. Indeed, many folk assume that the term 'black hole' itself originated with John Wheeler. Certainly, he adopted it in a lecture in 1967, when someone in the audience got fed up with him constantly referring to 'gravitationally completely collapsed objects' and asked why he didn't just call them black holes. That brought the term into common usage, but its origin has recently been traced back to another legendary physicist, Robert Dicke, at the start of the 1960s. And it first appeared in print early in 1964.

Notwithstanding its etymology, the idea of black holes has been around for much longer. You might be surprised to hear that it was an 18th-century English clergyman who first suggested that some stars could be so massive that not even light would be able to escape from them. Thus, they would be invisible. John Michell was an extraordinarily gifted thinker who, after a spectacular academic career in Cambridge, became rector of a church at Thornhill, a village in West Yorkshire, in 1767. There, he conducted mathematical investigations of many aspects of astronomy, gravitation, geology and other scientific pursuits – all with great originality. Michell published his work on what he called 'dark stars' in no less a journal than the *Philosophical Transactions of the Royal*

Society in 1783. His paper included the prescient suggestion that such objects might be discovered by looking for stars that seemed to be orbiting around – absolutely nothing. And that's pretty well how some black holes are discovered today.

MICHELL WAS SO FAR AHEAD OF HIS TIME THAT HIS WORK was essentially ignored and, like the contributions of a handful of other early visionaries, soon forgotten. And there the idea of dark stars lay until early in the 20th century, when another genius by the name of Karl Schwarzschild moulded Albert Einstein's newly minted theory of gravity into a form that unambiguously predicted their existence. Einstein's theory, usually known as the General Theory of Relativity, says that anything with mass distorts the space and time around it, and we feel that distortion as gravity. As does any other massive object, as it slides along the distortions in response.

The idea of empty space warping weirdly as a result of the presence of matter is counterintuitive, as is – even more so perhaps – the idea of the same thing happening with time. But since Einstein's theory was published late in 1915, it has been tested to within a nanometre of its life, and has come through each time with flying colours. And, today, it has everyday practical applications. GPS, for example, simply wouldn't work if it didn't take general relativity into account. Schwarzschild's research in the aftermath of Einstein's publication had far fewer immediate consequences, but it laid down the theoretical basis of black holes. He looked at the way gravity behaves in the vicinity of a spherical lump of matter and, indeed, the mathematical solutions he developed work well for run-of-the-mill celestial objects like planets and stars. But, crucially, they also predict what happens if that spherical lump of matter is shrunk to an infinitesimal point.

What they tell you is that the infinitesimal lump gives rise to an imaginary sphere from within which nothing can escape – not even light. The sphere is centred on the lump, and its radius is a distance that we now call the Schwarzschild radius. As I hinted a few paragraphs ago, that sphere is known as the event horizon. While you, as an intrepid space traveller, could easily fall through it without noticing, an observer watching from a distance would see you approach it ever more sluggishly as a result of the way gravity slows down time (technically termed gravitational time dilation). Eventually, to the outside observer, you would appear frozen on the event horizon – never seeming to cross it. Sadly, however, although you would not have noticed it going by, things would soon start to go pear-shaped for you. Well, that's actually a very poor metaphor, because as you approached the infinitesimal lump, your feet would feel more gravity than your head, and you would be drawn uncomfortably into something long and thin – a process known (even in the trade) as 'spaghettification'.

While an 'infinitesimal lump' is certainly what constitutes a black hole, we normally use slightly different language to describe it. It's known as a singularity – a single point where the density of space is infinite. I'll return to that crazy idea shortly, but for now note only that infinite density doesn't mean infinite mass. In fact, the amount of matter in a black hole is what defines its Schwarzschild radius – that is, the radius of the event horizon. More mass means a bigger event horizon, so, for example, a black hole with the mass of the Sun (which would be known as a one-solar-mass black hole) would have an event horizon about 6 kilometres in diameter, while the one for a black hole with the mass of Earth would be only 18 millimetres across – the size of a small coin. Rather puts us in our place, doesn't it?

IT'S UNLIKELY THAT SCHWARZSCHILD EVER IMAGINED that real celestial objects would be discovered with the characteristics predicted by his solutions of Einstein's general relativity equations. And, sadly, his life was cut short only months after he developed them. He died in May 1916 at the height of the First World War, while serving in the German Army on the Russian front. He suffered from a rare autoimmune condition known as pemphigus, which eventually took his life. He was 42.

Back in the world of astronomy, theoretical physicists began to postulate objects with successively more peculiar properties. We know that normal stars are held in a delicate balance between gravity and the pressure of radiation generated in their centres by nuclear reactions taking place there. Throughout their lifetimes, they undergo several distinct stages as the hydrogen fuel that powers their nuclear furnaces becomes depleted. But what happens when the fuel runs out altogether? Basically, gravity wins, and their cores collapse. And the end product of the collapse is something that depends on the original mass of the star.

During the 1930s, astronomers recognised that the end product of the collapse in most normal stars is an object called a white dwarf, which is in a condition rather unflatteringly described as 'electron degenerate'. It sounds decidedly suspect, but it means that the star's collapse under its own gravity has been halted by the pressure of electrons jostling together. What you wind up with is an object with the mass of the Sun compressed into the diameter of Earth. Many examples of white dwarf stars are known, and, indeed, it's the fate in store for our Sun when it runs out of fuel in about five billion years' time.

If you have a collapsing star with more than 1.4 times the mass of the Sun, however, the electron pressure won't stop the rot, and gravity will keep on compressing it until something else stops it. Shortly before the Second World War, scientists realised that this something else is the pressure of neutrons jostling together. That will produce an object exhibiting neutron degeneracy. Think stars like the Sun compressed into something the size of a city. But in a star with more than about 2.2 times the mass of the Sun (a limit recently confirmed using gravitational waves, which I'll explain in more detail in chapter 20), neutrons won't halt the collapse. And neither will anything else. So the collapse just continues into a singularity: a point in space with zero dimensions – otherwise known as a black hole.

A major finding by someone I know quite well, having been lucky enough to work alongside her in Edinburgh in the 1980s, spurred astronomers on to accept the reality of black holes. This was the discovery of pulsars, made in 1967 by Dame Jocelyn Bell Burnell, then at Cambridge University. As we noted in the last chapter, pulsars are objects that emit brief pulses of radiation with incredible regularity, and, by 1969, they had been recognised as rapidly spinning neutron stars beaming out radiation along their magnetic poles. So, if gravitationally collapsed neutron stars were a reality, could black holes be, too? Then, in 1971, the first likely candidate for black-holeship was identified – an X-ray source in the northern-hemisphere constellation of Cygnus, imaginatively named Cygnus X-1. In those days, X-ray astronomy was in its infancy, since it can only be carried out from above Earth's atmosphere. Now, however, with the combined strength of a brigade of modern X-ray satellites and a suite of radio and optical telescopes on the ground, we have a very good idea of Cygnus X-1's vital statistics.

SITUATED 6070 LIGHT YEARS FROM THE SOLAR SYSTEM, Cygnus X-1 consists of a 15-solar-mass black hole being orbited by a giant star, from which it is leaching gas. That gas falls into a swirling disc of material around the black hole's equator (known as the accretion disc), whose violent motion is the source of the X-ray and radio emission. Moreover, the black hole itself is spinning at around 800 revolutions per second, generating colossal magnetic fields that focus two jets of fast-moving material outwards along the rotation axis of the black hole.

If you could stand close to Cygnus X-1, the swirling accretion disc and the two jets at right angles to it would dominate your view. The event horizon embedded in the middle would look black, of course, as it does in the new image of the Messier 87 black hole. But around it, space is so tightly curved by the black hole's gravity that it acts like a strong lens, giving you a highly distorted view of the accretion disc behind it. And if you were near enough to it, you'd be able to see the back of your own head – but trying that is not recommended.

The characteristics of Cygnus X-1 are typical of what are known as 'stellar-mass' black holes – ones that have roughly the mass of a single star (although Cygnus X-1 is rather on the beefy side for this category). Something like 30 are known throughout the Milky Way Galaxy, but astronomers guess that there are many more lurking out there – perhaps even many millions more.

But that total is well and truly eclipsed by a different class of black hole, of which we have now discovered well over 70. And these are the monsters of the Universe – the so-called supermassive black holes, whose sizes are measured not in solar masses, but in millions of solar masses. It is suspected that all galaxies have a supermassive black hole at their centres, with the largest being measured in tens of *billions* of solar masses.

How do you weigh a supermassive black hole? The usual method relies on the black hole's ability to heat gas as it falls in towards the object. Some subtle thermodynamics tells you that the higher the temperature of the gas, the bigger is the black hole.

Even for distant galaxies, gas temperature can be gauged by the energy of the X-rays that it emits, so we can indirectly measure the central black hole's mass. In the case of our own Milky Way Galaxy, however, the black hole is on our cosmic doorstep at a distance of some 26 000 light years, so our measurements can be much more direct, and consequently more accurate.

Over the past 20 years or so, at least two specialist teams of astronomers have been using large optical telescopes to peer through the smoky murk that obscures our view of our Galaxy's centre. They use the dust-penetrating power of infrared radiation, and have been able to plot very accurately over time the motions of a swarm of stars that appear to circulate around nothing. But here's the trick – you don't even need to be able to see what they're orbiting around to use the stars' velocities to determine how much it weighs. The answer they get is that the stars are circulating around something with a mass of 4.1 million times that of the Sun. While the black hole itself is invisible in infrared radiation, its accretion disc is a strong emitter of radio waves, which, incidentally, give the object its official name of Sagittarius A* (pronounced 'A-star'). Astronomers can use the radio emission – as well as the orbits of the swarming stars – to determine the maximum size of what is in the middle. And it's so compact that it can't be anything other than a black hole.

While it is more than a thousand times less massive than the behemoth at the centre of Messier 87, the Sagittarius A* black hole is much nearer, and makes an obvious target for the Event Horizon Telescope. In fact, as the world was told at the April 2019 media conference, it has already been observed and the data are being reduced. At the time of writing, however, the results haven't been released.

HOW HAVE THESE SUPERMASSIVE BLACK HOLES GROWN so big? Our best guess is that it is by consuming gas and stars from the central regions of their host galaxies, and several detailed mechanisms have been proposed for this. Despite the sometimes slow rate of accretion, the lifetimes of galaxies are very long, allowing the black hole to grow steadily. There's also evidence from what is known as quasar activity (in which the compact nucleus of a galaxy becomes extremely bright for a limited period) that in certain early stages in a galaxy's evolution, the central black hole consumes vast amounts of gas and dust, causing an enormous increase in the luminosity of its surrounding accretion disc and accompanying jets. By looking deep into intergalactic space, and thus into the distant past, we can make detailed observations of these quasars, improving our understanding of the physical processes that drive them. In today's Universe, they are extinct.

One remaining puzzle in our understanding of black holes is that there's a curious size gap. We have stellar-mass black holes, and we have supermassive black holes, but there doesn't seem to be anything in between. Searches for these so-called intermediate-mass black holes – with masses a few hundred times that of the Sun – have turned up a few, but the identifications are in many cases not secure. Some candidates are close to the centre of our own Milky Way Galaxy, and others are embedded in ancient globular clusters of stars that orbit around our Galaxy and may themselves be remnants of dwarf galaxies that have been torn to shreds by its cannibalism. Much work on this topic is underway, and you can be sure there will be more results in coming years.

THERE'S ONE MORE IMPORTANT POINT I'D LIKE TO MAKE about black holes, and it links two things that might at first seem unrelated. One is what I've alluded to already: the idea of a singularity – a single point in space that has infinite density. Density is just mass divided by volume, and it's the fact that a single point has zero volume that propels the density to infinity, no matter what the mass might happen to be. A point with zero volume is very hard to get your head around, even if you happen to be an astrophysicist. You won't have to look too far online to find an entertaining video clip of a gaggle of expert Australian astronomers trying to explain what a black hole is. Most of them wind up throwing up their hands in exasperation – because we simply don't have the language to explain what's going on in the singularity.

Perhaps a clue comes from the similarly counterintuitive idea that black holes can evaporate. 'But wait,' I hear you cry, 'I thought you said that nothing can escape a black hole?' Yes, I did, and it's true that our best theory of the way gravity works – general relativity – says it can't. But it was the late great Stephen Hawking who, in 1974, postulated theoretically that black holes can lose mass by emitting something now known as Hawking radiation. It's electromagnetic radiation that occurs over a very broad range of wavelengths – but is exceedingly weak. So weak, in fact, that despite being universally accepted as a reality, it has never been experimentally confirmed.

How does it happen? Perhaps the simplest way to imagine how Hawking radiation is created is to think of entities known as virtual particle pairs, which quantum physics tells us pop into and out of existence in a vacuum. When that happens close to a black hole, some of the pairs of virtual particles can get separated by the event horizon, with one particle being stuck forever inside the black hole, and the other flitting off into space. You'll have to trust me that the net effect of this is that the black hole loses mass, albeit at a vanishingly small rate. But the mass loss is why it's termed 'evaporation', and it turns out that the smallest black holes evaporate most quickly. Even then, however, the timescales are much,

much longer than the present age of the Universe. For example, the number of years it would take a black hole of one solar mass to evaporate is 10 followed by 64 zeroes. Talk about watching paint dry.

The point I want to make in all this is that relativity doesn't provide all the answers. At some level – in the singularity, perhaps – the physics brings in quantum mechanics, the weird but also extremely robust theory of the way things behave on very small scales. And at present, we have no unified theory that links the two. Despite the best efforts of physicists worldwide, we have no confirmed theory of quantum gravity, and that means our understanding is incomplete. In some ways that's very exciting, because the 'new physics' that might eventually reconcile quantum mechanics and relativity could lead us into dimensions beyond time and space, and a deeper understanding of cosmic mysteries like dark matter.

Stephen Hawking passed away on 14 March 2018 (Einstein's 139th birthday), and is much missed in the world of physics. I never met him formally, although he did once run over me in his wheelchair – an honour I think I share with several other inhabitants of Cambridge, where he was a familiar figure. As I write these words, a new British 50 pence coin is entering circulation commemorating Stephen's life. It carries on it a clever depiction of a black hole, together with the equation for black hole entropy (its degree of disorder) that Stephen Hawking and Jacob Bekenstein arrived at in the early 1970s. It provides a tangible reminder that in black-hole physics, we still have much to learn.

CHAPTER 19
THROUGH GRAVITY'S LENS: THE CURIOUS MATTER OF DARK MATTER

A black hole is the most extreme example of gravitationally warped space, entrapping light within its event horizon. But in less daunting environments, warped space can be a really useful tool for astronomers – for it turns out that it can act like a gigantic natural telescope.

You won't be surprised to hear that the idea goes back to the early 20th century, when Albert Einstein was working on the General Theory of Relativity that revolutionised our understanding of gravity. Our understanding hadn't changed much since 1687, when Isaac Newton had set out his own Law of Universal Gravitation. Newton's theory says that gravity is a force that makes any object in the Universe attract any other object. Happily, it works perfectly well for normal objects like stars, planets, humans and small animals. But, as Einstein noticed with the planet Mercury, wherever gravity is strong, Newton's law breaks down.

Einstein's theory was published in 1915, and doesn't miss a beat in strong gravity because it dispenses altogether with the idea of forces between individual objects. Gravity, it says, is a property of the Universe as a whole. General relativity sees gravity as a distortion of the underlying 'fabric' of space (or, more accurately, of space-time, since, in this context, time acts like a fourth dimension). The distortion is caused by the presence of matter, and one of the theory's first predictions was that light passing close to a massive object like a star will be bent by a tiny angle as it travels through the distorted space.

In the case of the Sun, Einstein predicted the bending would amount to 1.75 seconds of arc. For an idea of what this means, imagine one of the new British 50 pence coins commemorating Stephen Hawking (which are 27.5 millimetres across – a little more than an inch) held up at a distance of just over 3 kilometres, or 2 miles. The disc of the coin covers an angle of 1.75 seconds of arc, and would be totally invisible to you without a sizeable telescope. But astronomers have ways of measuring such minuscule angles in a branch of the science known as astrometry.

And so it was that on 29 May 1919, the British astrophysicist Arthur Eddington took advantage of a rare alignment of the Sun and Moon to measure the positions of several distant stars close to the disc of the Sun. Fortuitously, the total solar eclipse he photographed that day occurred right in front of a rich cluster of bright stars known as the Hyades, allowing him to measure the extent to which they appeared to be deflected. Eddington's results confirmed Einstein's prediction, which immediately rocketed the physicist to world fame, and – perhaps more importantly – went a long way towards healing the bitter scars of war. They had been particularly acute in the scientific community, but now, a British astronomer had confirmed the theoretical prediction of a German-born physicist – and both were ardent pacifists. It was good news.

BACK IN 1912, WHILE STILL REFINING HIS THEORY, EINSTEIN had visited an astronomer colleague at the Berlin Observatory by the name of Erwin Freundlich. (In fact, Freundlich was my 'academic grandfather'. Decades later, he moved to Scotland where he trained a young Polish astronomer named Tadeusz Slebarski, who subsequently became my masters degree supervisor. I never met Freundlich, but by all accounts, he was a delightful man who wore his brilliance lightly. As, indeed, was Slebarski.)

Einstein and Freundlich became great friends, sharing a love of music as well as science. During their 1912 get-together, they investigated one of the more esoteric consequences of relativity – that a massive object in space would have an unusual focusing effect on light from a more distant source behind it. What you would see from an earthly vantage point would depend on how perfectly aligned the massive object and distant light source were. Normally, because the alignment wouldn't be exact, the distortion induced by the nearby massive object would create two images of the more distant source – one on each side of the nearer object, which, despite being massive, might be so faint as to be invisible.

While Einstein and Freundlich didn't have the computational power to deal with it, we now know that the intrinsic shape of the nearer object (a galaxy, for example) would also affect the distortion of space around it. Under some circumstances, it could even produce four images of the distant source arranged in a cross – now known as an Einstein cross.

And what would happen if the distant object, the nearer one, and Earth lined up in a perfectly straight fashion? Einstein figured that out, too, but didn't write it up for publication until 1936. Such an exact alignment would result in the distant source appearing as a small open circle in the sky. We now call it an Einstein ring, although the man himself thought the probability of observing such an exact line-up in the real Universe was so small that it would never be anything more than a mathematical curiosity.

In a couple of rather brusque scientific papers written at the end of 1919, Oliver Lodge (a British physicist who didn't really believe in relativity) suggested that the distortion of space-time by a massive object would produce exactly the same effect on light as a glass lens does. It would be an unusual one, however, with a raised cusp at the centre rather than the smooth convex or concave surfaces of a normal lens. It turns out that he was right. And, over many years of experimentation, I've noticed that a decent approximation to such a cusped lens is formed by the base of your average wine glass.

IN THE EARLY 1960s, ASTRONOMERS BEGAN TO DISCOVER the remote, point-like objects known as quasars. We now know that they are the delinquent nuclei of young galaxies, powered by supermassive black holes. In the 1960s we had no idea what they were, but it was quickly discovered that they must be very distant. And, then, in 1979, a strange 'double quasar' was found – two quasars side by side, with uncannily similar properties. I well remember the excitement among my colleagues when it was suggested that this could be the first example of one of these mythical gravitational lenses. And, indeed, it was – a single distant quasar gravitationally lensed by an intervening normal galaxy that was too faint to show up in the telescopes of the time.

Not long after that, astronomers started discovering mysterious short arcs of light on photographic images of deep space. Once again, there was great excitement, although some proposed that they were just coffee-mug stains on the photographic images. A suggestion that they might be bits of spiral galaxies that had broken off was also quickly dismissed, as astronomers remembered gravitational lensing. And yes, the arcs were eventually confirmed to be incomplete Einstein rings. It was not until 1988 that the first full Einstein ring was found using radio telescopes, but we now know of many. There are also many examples of multiple images produced by gravitational lensing, including some spectacular Einstein crosses.

Even stars and planets have a gravitational lensing effect – called microlensing – which has allowed invisible planets around other stars to be detected from Earth. Typically, a dim, nearby star will pass in front of a more distant one. During its transit, the distorted lens-like space around the nearer star amplifies the light of the distant one, sometimes by as much as a thousand times. Plotting a graph of the light over the weeks or months of the transit reveals a peak in brightness at the time when the two stars are most closely aligned. If, as is often the case, there is one or more subsidiary peaks lasting only hours or days, they reveal the presence of planets around the nearer, lensing star.

These are one-off observations – the lensing effect, being the result of a chance alignment, is never seen again. Microlensing observations can still contribute much to our overall knowledge of extra-solar planets, however, particularly for low-mass planets orbiting at large distances from their parent stars. Other detection methods are pretty insensitive to such objects. It's for this reason that a number of aptly named international collaborations exist to monitor the sky for microlensing events – for example, OGLE (the Optical Gravitational Lensing Experiment) and MicroFUN (the Microlensing Follow-Up Network).

LET ME NOW GO BACK A FEW DECADES TO BRING ANOTHER strand into this story. Back in 1933, an eminent Swiss-American astronomer by the name of Fritz Zwicky made some puzzling observations of a cluster of galaxies. These occur commonly throughout the Universe, and may include hundreds or even thousands of galaxies concentrated in a relatively compact region of space.

The cluster that Zwicky was observing was large (more than 1000 galaxies) and relatively nearby (around 320 million light years away). It's located in the northern-hemisphere constellation of Coma Berenices, or Berenice's Hair (after an Egyptian queen who offered her tresses as a thank you to the gods). Using spectroscopic techniques, Zwicky was measuring the velocities of the individual galaxies in the cluster, and what he found surprised him. If the galaxies he could see represented the cluster's total content, there was a problem, because the individual motions of the galaxies were too great for the cluster's gravity to hold onto them. The cluster should therefore have evaporated billions of years ago, with the individual galaxies going their separate ways at high speed. Zwicky reasoned that there must be some invisible component in the cluster that was holding onto them by its gravity. He even gave it a name when he published his findings in 1937 – 'dark matter'.

I think it's fair to say that the astronomical world was so baffled by this discovery that they just ignored it. It wasn't until 1970 that the concept of dark matter raised its head again, when another eminent astronomer realised that something didn't make sense. This was Ken Freeman of the Australian National University in Canberra, an old friend and colleague of mine whose pivotal work in 1970 has only been properly recognised in recent decades. Rightly, he now holds Australia's highest scientific awards and honours. Ken studies galaxies in detail, and, during the late 1960s, was interested in the way they rotate. This can be investigated by choosing disc galaxies that are edge-on to our own. While the beautiful spiral structure of such galaxies is forever hidden from us, there's some compensation in the fact that it's relatively straightforward to measure their rotation velocities using a spectrograph. In 1970 he published the disquieting result that his galaxies were rotating too fast for gravity to hold them together. So fast that they should simply fly apart. But once again, the astronomical community just shuffled its feet and got on with other things that were easier to explain.

When one of the most remarkable people in astronomy found the same result eight years later, however, things changed. The world took the problem seriously at last. Vera Rubin was not only a truly gifted scientist and a much-loved figure in the world of astronomy, but she was also one of the early champions of women in science. For most of her career, she was based at the Carnegie Institution in Washington, D.C. Sadly, we lost Vera on Christmas Day, 2016, She was 88.

Working in the 1970s with a new spectrograph built by her collaborator, Kent Ford, Vera made detailed measurements of the rotation of galaxies. In fact, she plotted what are known as rotation curves – graphs showing how a galaxy's rotation speed changes with distance from its centre. If there was nothing more in the galaxy than the stars, dust and gas you can see, the rotation should be rapid near the middle, but should then fall away with increasing distance. Rubin and Ford found the opposite – the rotation velocity kept on increasing, gradually flattening off in the outer reaches of each galaxy. The most straightforward interpretation of this is that galaxies are embedded in spheroidal halos of something massive but completely invisible. Zwicky's dark matter again – which promptly became the hot topic in astrophysics, with extensive resources mustered to tackle the problem of its identity.

SCIENTISTS QUICKLY CONCENTRATED THEIR EFFORTS ON three possibilities. First was the idea that dark matter doesn't actually exist, and something else is wrong with our understanding of physics. The most serious attack along those lines came from Mordehai Milgrom, a physicist at the Weizmann Institute in Israel. Milgrom reasoned that at the very low accelerations experienced by stars orbiting within galaxies (which are much, much lower than those experienced by planets in the Solar System), they might obey different rules of physics. So he proposed something known as MOND, for MOdified Newtonian Dynamics, which was aimed at eliminating the need for dark matter.

Milgrom's theory was published in 1983, and, itself, immediately came under attack. The criticisms centre around the fact that while it might make sense of the internal motions of galaxies, in many other respects, it doesn't match astronomical observations. One failing, for example, is that it doesn't quite get rid of the need for dark matter altogether. Another, more fundamental one, is that MOND predicts that the speeds of light and gravity should differ, while, as we will see in the next chapter, they have now been measured to be identical. While Milgrom's theory is still being researched, mainstream astronomy has never accepted it as the solution to the dark matter problem.

So, if it's not MOND, what is left? There were two other possibilities for dark matter: MACHOs and WIMPs, perhaps the most appropriate acronyms in the whole of astrophysics. MACHOs are the bruisers. They are Massive Compact Halo Objects – invisible things like massive black holes with no surrounding accretion discs, dead white dwarf stars, orphan planets and other dark celestial detritus lurking in Vera Rubin's spherical halos around galaxies. We've known for over half a century that a substantial population of old, dim stars occupies a spherical region around most spiral galaxies, appearing much fainter than the galaxies' discs. So the idea of a halo is not new – it's what it might contain that's at issue. But – it's not MACHOs. They were effectively ruled out in the 1990s when astronomers observed large numbers of stars in search of the gravitational microlensing events that MACHOs would produce – and did not find them.

'BEACONS OF LIGHT ON HILLS OF DARK MATTER'

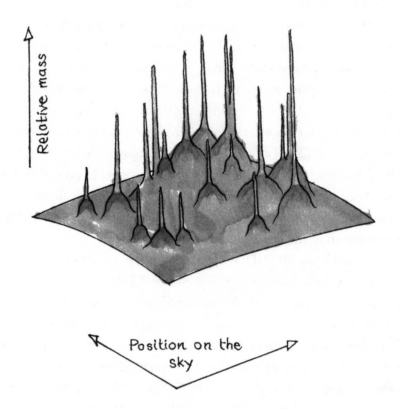

Not stalagmites on a cave floor, but the distribution of matter in a cluster of galaxies revealed by its gravitational effect on the light from more distant objects. As well as the spiky mass concentrations of the individual galaxies, the observations show the dark matter halos in which they are embedded.

Author, after LSST

Which leaves WIMPs – Weakly Interacting Massive Particles. Some species of subatomic particle that never (or very rarely) interacts with normal matter in any other way than gravitationally. Thus it is gravitational weirdness that has revealed their existence. These particles have not yet been discovered, but we know a lot about them, thanks to astronomical observations. We know, for example, that dark matter is, indeed, concentrated in halos around galaxies. In fact, we can map the exact distribution of dark matter within a cluster of galaxies by making further use of our old friend, gravitational lensing.

Since the advent of the Hubble Space Telescope in 1990, we have been able to study galaxy clusters in great detail. And many of them reveal multiple short arcs of light surrounding the cluster and centred on it. These are the distorted (but also amplified) images of far more distant galaxies behind the foreground cluster, gravitationally lensed by the matter in it – both visible and dark. By making statistical assumptions about the true shape and distribution of the distant galaxies, all the matter in the foreground cluster can be mapped – including the structure of the dark matter blob in which it is enveloped. One of my colleagues, Matthew Colless of the Australian National University, has eloquently described galaxies in clusters as 'beacons of light on hills of dark matter', and that is exactly what these maps reveal.

What is remarkable is that without the gravitational distortion provided by the foreground cluster, many of the background galaxies would be completely invisible to our telescopes because they are so remote. Truly, the cluster itself has become a gigantic natural telescope, hundreds of millions of light years across, revealing objects that by rights, we shouldn't be able to see.

THE BOTTOM LINE IN ALL THIS IS THAT DARK MATTER likes to be where normal matter is. That means that its particles are not only in the room where I'm typing these words, but they also surround you as you read them. Having no interaction with the normal matter of which we and everything around us is made, the gazillions of dark matter particles constantly passing through us are completely undetectable.

Not only that, dark matter outweighs normal matter by five to one overall. It's by far the largest material component of the Universe, a measurement that comes not only from studies of individual galaxies, but also from large-scale galaxy mapping surveys. Looking at that overall big picture, we can infer subtleties like the five-to-one ratio of dark to normal matter. In fact, when you feed all this into models of how the Universe has evolved in its 13.8 billion-year history, it turns out that but for the dark stuff, we wouldn't be here. In the aftermath of the Big Bang, the Universe consisted of a web of dark matter, which provided a kind of gravitational framework around which hydrogen concentrated to form stars and galaxies – and eventually planets and us. We see the remaining evidence of that framework today in the large-scale distribution of galaxies across the Universe.

I think it's fair to say that with a few exceptions (possible mutual annihilation of dark matter particles concentrated in the centres of galaxies, for example), astronomy has run its course with dark matter, and the campaign to discover its true nature has passed to the particle physicists. They are thrilled, of course, because several different aspects of the subatomic world already point to an incompleteness in their picture of what makes up the Universe around us – the so-called Standard Model of Particle Physics. Dark matter only serves to confirm that, giving them something else to chase.

Particle physicists have evolved a complex theoretical model that copes with some of the incompleteness, and also includes candidates for dark matter. It's called supersymmetry, or susy for short, and postulates that the standard particles, such as electrons, muons, quarks and the like, have a supersymmetric 'shadow particle' of much higher mass. Where are these shadows supposed to be lurking? In hidden dimensions? Maybe – I confess I'm not an expert on supersymmetric particles. Actually, I'm a complete amateur. But I do share the disappointment of my colleagues at major experimental facilities like the Large Hadron Collider – the giant atom-smasher that straddles the French–Swiss border near Geneva – that so far, no trace of susy has been seen. Many particle physicists are now wondering if we're barking up the wrong tree, and that some other theoretical framework is necessary.

I've been privileged to make several visits to CERN, the European Nuclear Research Centre where the Large Hadron Collider is located. It's a wonderland for physicists, with a leafy outdoor cafeteria that, on summer lunchtimes, is full of scientists discussing their research. But it also boasts something else that I think puts everything into perspective. Near the edge of the cafeteria is a grassy area with a small rabbit hutch. And on it is a sign that reads 'CERN Animal Shelter for Computer Mice'. And sure enough, when you look inside, there are dozens of elderly computer mice with plenty of food, water, straw and shelter to keep them happy and content in their retirement. And, of course, with complete protection from marauding computer cats.

As you might have guessed, I admire scientists who work hard at their research but don't take themselves too seriously. While I can do little more than cheer these particle physicists on from the sidelines of their research, I feel confident that it won't be too long before we've cracked the dark matter nut.

CHAPTER 20
RIPPLES IN SPACE: PROBING THE BIRTH OF THE UNIVERSE

There is one space-related date in recent years whose importance is almost impossible to overstate. We have been fortunate in our era to have witnessed the opening of a new window on the Universe, one whose potential is truly breathtaking. As is the engineering that has facilitated it. I'm talking about the first detection of gravitational waves by an extraordinary 'telescope' in the United States on 14 September 2015. The news was not released until the following February, but the science press immediately grasped its significance, proclaiming the discovery with breathless enthusiasm. The discovery was made with an instrument called LIGO – the Laser Interferometer Gravitational-Wave Observatory – which is a gobbledygook way of describing a machine that senses minute changes in length.

What do I mean by minute? LIGO can detect changes in the length of a 4-kilometre-long beam of light amounting to one ten-thousandth of the diameter of a proton – the subatomic particle at the centre of a hydrogen atom. In maths-speak, that's precision of better than 10^{-19} metres.

While your head is reeling from that, as mine did when I first learned about it, let me explain how LIGO's name gives away the technology. Technically, it's known as 'Advanced LIGO' to distinguish it from earlier development versions, but that needn't worry us. The light beam I just mentioned comes from the laser, of course. What's an interferometer? This is a device that takes a beam of light, splits it in two, and then recombines it so that the individual light waves come together in step – or very nearly so. If two waves are wildly out of step when they recombine, with the peak of one hitting a trough of the other, for example, they'll cancel out altogether and produce darkness – a quirk of physics that I've always regarded as slightly magical. But two waves that come together just a fraction out of step can be compared and measured with high accuracy in an interferometer, which is why it's such a powerful instrument.

And what's the point of splitting the light in the first place? This is done so that the two resulting beams are at 90 degrees to each other. If one beam was sent off to the north, for example, the other would be directed west. In fact, LIGO is at a rather different compass bearing, but its two 4-kilometre-long arms are still exactly at right angles, like a gigantic 'L'. At each tip of the 'L' are mirrors that send the light beams back to be recombined. So, what LIGO is sensing with that head-spinning precision is the minuscule discrepancy in length between two identical light beams that differ only in their direction in space.

And that's where the gravitational-wave bit comes in. We're not talking about light waves now, nor seismic waves rattling the surface of Earth, but ripples in space-time itself. Their existence is a consequence of Einstein's General Theory of Relativity, which, you will recall, says that space is not perfectly rigid, but flexes minutely in response to matter. Things that can flex tend to be able to transmit waves, too – think of sound waves in air, ripples on a pond, heavy metal on guitar strings, and so on.

With Einstein's general relativity withstanding every critical test that has been thrown at it over the past hundred-odd years, the prediction of gravitational waves was always taken seriously. And it had long been anticipated that when equipment could be designed and built with the required level of sensitivity, it would be able to detect them. Now, after decades of development, this has come to pass, with minute tremors in the length of LIGO's light beams revealing the passage of a gravitational wave. Curiously, the waves detected fall within the range of frequencies to which the human ear is sensitive, so when the LIGO signals are hugely amplified, you can hear them.

I've rather glossed over this description of LIGO, neglecting, for example, the fact that there are two of these large interferometers – one at Livingston, Louisiana, and the other at Hanford, Washington. If you look at those two places on a map, you'll see that they are at opposite corners of mainland United States – at least as far as local geography will allow. If you could shine a flashlamp from one to the other, the light would traverse the distance in about 10 milliseconds, or one-hundredth of a second. That's important because theory predicts that gravitational waves travel at the speed of light, so having two widely spaced detectors gives you a handle on what direction they're approaching from.

SO, WHAT CAUSES GRAVITATIONAL WAVES? THEY ARE emitted when any massive object is accelerated. In the case of that first detection in 2015, the waves originated in a major gravitational disturbance some 1.3 billion years ago, when two distant black holes spiralled together and merged to form a bigger one. Such a merging takes place over time, with the black holes revolving ever more rapidly around each other during the last few seconds. The resulting gravitational waves pulse outwards with increasing intensity and frequency as the two black holes spin together – but vanish once they have merged, in what's called the 'ring-down' of the resulting black hole. Played in audio, the waves sound like a short whistle whose loudness and pitch increase ever more rapidly before suddenly falling silent – a so-called 'chirp'.

You probably won't be surprised to read that the first gravitational-wave detection on 14 September 2015 is designated GW150914. But you might be more impressed to know that the analysis of its signal yielded not only the distance of the event (1.3 billion light years) but also the masses of the two merging black holes (35.6 and 30.6 times the mass of the Sun) plus the final black hole mass of 63.1 solar masses. Note that those mass figures don't add up, and what the discrepancy tells you is that the merger also produced the energy equivalent of 3.1 solar masses in gravitational waves. (You can work out how much energy that is by using a rather famous equation that links it with mass and c, the speed of light, squared.) The colossal energy release is why those ripples in space were still detectable after more than a billion years of travelling through the Universe.

As might be expected, that first detection yielded a handsome return in international awards, with the 2017 Nobel Prize for Physics going to Rainer Weiss, Kip Thorne and Barry Barish for their role in it. And since then, significant progress has been made in gravitational-wave astronomy. Advanced LIGO has been joined by a European instrument known as Advanced Virgo, located near Galileo's old stamping ground in Pisa. Others are in development. The benefit of adding to the world's suite of gravitational-wave observatories is that more well-spaced detectors enhance not only our sensitivity, but also our ability to pinpoint the direction from which a given signal has come.

At the time of writing, GW150914 has been joined by ten further confirmed event detections with many more in the pipeline (one of which, reported in May 2019, may be the first example of a black hole devouring a neutron star). Of the ten confirmed, just one – GW170817 – did not come from merging black holes, but from merging neutron stars at a distance of about 130 million light years. The gravitational-wave signal came from the final 100 seconds of this event, and once again, the chirp revealed the masses of the progenitor objects – 1.5 and 1.3 times the mass of the Sun. But while a merger of stellar-mass black holes is not expected to produce an electromagnetic pulse, a merger of neutron stars is. And, in a triumph of international collaboration, GW170817 was detected by 70 observatories worldwide (and in orbit) across the entire electromagnetic spectrum. The detections ranged from gamma rays, which arrived 1.7 seconds after the gravitational-wave signal, to radio waves detected 16 days later. When the physics of the various emission mechanisms was taken into account, these observations provided spectacular confirmation that gravitational waves do, indeed, travel at the speed of light.

I WANT TO TURN, NOW, TO A PICTURE THAT IS EVEN bigger than mergers of exotic objects in deep space. The generally accepted theory of the beginning of the Universe is something called the Big Bang, which postulates an origin of everything (space, time and matter) some 13.8 billion years ago. It's not the only theory espoused by the science of cosmology – which studies the origin and evolution of the Universe as a whole – but it's the one with the most solid evidence. It's based on a mixture of observation and general relativity, which forms its theoretical foundation. While 'Big Bang' is an evocative description, cosmologists often refer to a more precise mathematical formulation known as the 'Lambda CDM model', which I'll explain in a few minutes. Other theories, such as a regenerative Universe, or even multiple ones, are more speculative, even though they're great to talk about at parties. Well, the kind of parties I go to, anyway.

So what is the Big Bang theory? It owes its origin to the work of a Russian mathematician named Alexander Friedmann, and a Belgian priest called Georges Lemaître, who independently formulated some of the properties of an expanding Universe in the 1920s – before Edwin Hubble discovered the actual expansion in 1929. Lemaître, in particular, focused on the idea of a 'primaeval atom' from which the Universe and its contents have evolved. Over the next few decades, the theory was refined to include the idea of an extremely hot and dense beginning, which, today is regarded as a singularity – but with infinite temperature as well as infinite density. Present-day physics is not equipped to probe the innards of this singularity, but provides a surprisingly good understanding of its immediate aftermath.

IN THE MIDDLE YEARS OF THE 20TH CENTURY, THE BIG Bang theory was pitted against a competing model that envisaged matter as being continuously created within an infinitely old Universe. That was the 'steady-state' theory, espoused by British astronomer Fred Hoyle and others. But two discoveries knocked the steady-state theory on its head. The first came from a new generation of sensitive radio telescopes that were introduced in the mid-1960s. One of them revealed a mysterious background hiss in the microwave spectrum that seemed to cover the entire sky. It took a while before scientists realised that what they were picking up was something that had been predicted nearly two decades earlier, in 1948. It was effectively the afterglow of the Big Bang – the brilliant light that had filled the infant Universe, stretched in wavelength into microwaves by its subsequent expansion. We give this afterglow a technical name – it's called the Cosmic Microwave Background Radiation, or CMBR.

Why can we still perceive this ancient fossil radiation? Once again, it arises because whenever we look into space, we are always looking back in time. In fact, once you get beyond the confines of our own Milky Way Galaxy, the so-called 'look-back time' becomes a more relevant concept than the actual distance. So, while the eight-minute look-back time to the Sun, or the 4.3-year look-back time to the nearest bright star (Alpha Centauri), or even the 2.5-million-year look-back time to the nearest big galaxy (Andromeda) aren't particularly momentous in evolutionary terms, once you get to more distant galaxies, you're looking back to a significantly earlier epoch. Which, incidentally, was the second blow that took out the steady-state theory − because astronomers could see that at look-back times of several billion years, galaxies were considerably different from today's galaxies, implying that they have undergone evolutionary changes. That would not be the case in a steady-state Universe.

But back to the CMBR. To understand its origin, you have to appreciate that for the first few hundred thousand years after the Universe came into being, it was filled with a fog of brilliant radiation. It was essentially a fireball. As with a fog of water droplets here on Earth, there was no way of seeing through it. Water droplets scatter light, and, in a sense, so did the radiation permeating the cosmos. But then, some 380 000 years after the Big Bang, the fog cleared fairly rapidly throughout the whole Universe, rendering it transparent, as it is today. So, as we look further and further back in time through our transparent Universe, we eventually come to the instant when the fog cleared, and can see it as a wall of radiation covering the whole sky. We call it the 'last scattering surface', and it's the fact that the Universe has expanded by around 1300 times since it became transparent that has stretched the radiation into microwaves. Were it not for that, the sky would be an encompassing sphere of brilliant light, and there would be no such thing as night.

A moment's thought will show that this sphere of microwave radiation (the last scattering surface) is receding from us at the speed of light – because the moment when the fog cleared is retreating second by second into our past. The best way to get your head around this is to imagine yourself in the Universe at the moment the dazzling fog cleared. Knowing that it cleared everywhere exactly simultaneously (in our thought experiment, at least), do you immediately see darkness? The answer is no, because one second after it clears, you'll be seeing a wall of illumination 300 000 kilometres away whose light has only just reached you. That brilliant wall will be 600 000 kilometres away after two seconds ... and so on. Right from the start, the flash of the Big Bang is receding into your past – and it still is.

Effectively, the CMBR is a gigantic optical illusion, because it's not a physical barrier. Enclosed within it, however, is everything we can see in the Universe. And it has another attribute, too, which is of immense importance in studies of the Universe's evolution. As the waveband of the CMBR has been changed by the expansion of the Universe from visible light to microwaves, so has its temperature, falling from several thousand degrees when it was emitted to 2.7 degrees above absolute zero today. Effectively, that's the temperature of space. And it's almost uniform over the whole sky. But not quite – the CMBR has ripples of temperature in it, at the minute level of about one part in 100 000.

Remarkably, those ripples originated in sound waves reverberating through the primordial fireball – the sound of the Big Bang, if you like. Despite the vanishingly small range of temperatures they cover, the ripples have been mapped in detail by a succession of space-borne radio telescopes over the past two decades, revealing much about conditions in the hot early Universe. They also provide a baseline for our investigations of the way the Universe has evolved, because the slightly cooler spots are regions of higher density in the fireball. They are thought to have been the seeds of the large-scale structure we see in the Universe today, as revealed by the way galaxies are distributed in space. Comparing today's Universe with the CMBR tells us about the details of the expansion, including, for example, the contribution of dark matter.

Maps of the CMBR are, like gravitational waves, Nobel Prize material, and form a vital basis for contemporary cosmology. Because they conventionally depict the CMBR in a range of colours, and because this radiation is behind everything else we can see in the Universe, I sometimes refer to it as the 'cosmic wallpaper'. But, vital though it is, the cosmic wallpaper has a downside. It forms a horizon beyond which we can never see with radio telescopes, visible-light telescopes, or any other kind that depends on electromagnetic radiation. It is impenetrable. But why should we want to see beyond it? The answer to that lies in the fact that 'beyond' in this context means 'earlier'. That is, if there was some means of penetrating the cosmic microwave background, we could detect events that took place *before* the Universe became transparent. And this would allow us to probe details of the Big Bang that, at present, are only in the realm of theory.

WHAT ARE THE KINDS OF THINGS WE'D LIKE TO KNOW about? One concerns the origin of something I've hardly dared touch on in this book. It's another 'dark' mystery, next to which the mystery of dark matter pales into insignificance. While our ongoing quest into the nature of dark matter has, at least, some possibility of success, this one stubbornly defies the efforts of theoretical physicists. And it's to do with the expansion of space itself.

In the 1970s and 80s, most cosmologists assumed that there would be enough matter in the Universe to gradually slow down its expansion by the mutual gravitational attraction of everything in it. Perhaps even to the extent that at some time in the distant future, the expansion might stop and turn into a contraction, with the ultimate fate of the Universe being a 'big crunch' as it collapsed back into a singularity. Nobel laureate Brian Schmidt (today the Vice-Chancellor of the Australian National University) famously referred to this reversal of the Big Bang at the end of the Universe as the 'Gnab Gib'.

During the 1990s, Schmidt was leading one of two groups of scientists that were independently using observations of distant supernovae to chart this expected deceleration of the Universe. But what they found was the reverse. To the astonishment of everyone, both groups discovered that for the past six billion years or so, the expansion of the Universe has been accelerating. Announced in 1998, that was the discovery that earned Schmidt the 2011 Nobel Prize in Physics, along with his colleague Adam Riess, and Saul Perlmutter, leader of the other group.

Of course, the immediate question was 'Why?' The supernova work, together with comparisons of the Universe's present large-scale structure with that in the CMBR, suggest that space itself is endowed with a pressure that's causing the acceleration. For want of a better term, we call it dark energy, and it's a property of the Universe as a whole rather than a local effect. While accelerating expansion is definitely the situation today, it may not always have been. We believe that during the first six or seven billion years of the Universe's history, the matter in it was sufficiently closely packed that the braking effect of its mutual gravity was enough to decelerate it. The acceleration kicked in only when galaxies were far enough apart for dark energy to begin to overcome gravity.

The most recent work suggests that the dark energy of a portion of space is related to its volume, so, as space expands, it gets more energetic and further accelerates the expansion. In that respect, it resembles a mathematical entity that Einstein introduced into his relativity equations in 1917, which he called the 'cosmological constant'. He denoted this constant by the Greek symbol lambda, which is why the name 'Lambda CDM model' is used. Lambda represents dark energy, and CDM stands for 'cold dark matter'. Those two entities, together with normal matter (which, for the record, we call baryonic matter in the trade, and is dominated by hydrogen) make up the mass-energy budget of the Universe. The best observational determinations have them in the ratio 68:27:5 (dark energy to dark matter to normal matter). Once again, the Universe has put us in our place by the fact that everything we can see only amounts to a measly 5 per cent of its contents. And it puts astronomers firmly in their place with the admission that we don't understand what makes up the other 95 per cent.

PHYSICISTS ARE CURRENTLY WORKING HARD TO TIE DOWN dark energy. More observations of the Universe's rate of expansion at different epochs in the past might give us more insights, and that is being accomplished using fibre optics technology. It could be assisted by gravitational-wave observations of distant neutron star mergers. They could be used to better calibrate our standard light sources, for example, improving the cosmic distance scale. But wouldn't it be wonderful if we could probe beyond the CMBR to see what was going on in the works while the Universe was still glowing brilliantly? That might seem a bit fanciful, but there is at least one more pressing reason why we'd love to achieve that.

It's one of the fervent hopes for the bright new future offered by the detection of gravitational waves. Remember, they're not just emitted when black holes or neutron stars collide, but when any massive object is accelerated. And we believe the granddaddy of all massive accelerations occurred a tiny fraction of a second after the Big Bang itself, when the whole Universe underwent a fleetingly brief episode of violent expansion. There are good reasons to believe that the infant Universe expanded by at least 10^{26} times (that's a 1 followed by 26 zeroes) when it was about 10^{-36} of a second old (and that's a 1 preceded by 35 zeroes and a decimal point). Yes, I know these numbers seem ridiculous. They mean that instantaneously after the Big Bang, the Universe went from being the diameter of a hair to the diameter of a galaxy. And, with consummate understatement, we call this the period of inflation. It was followed immediately afterwards by the much gentler expansion that is still taking place today.

The inflation theory was developed in the late 1970s in order to overcome some of the problems of the Big Bang model as it was then understood. The almost perfect smoothness of the CMBR's temperature was one of them, because physics wouldn't allow the fireball to achieve such a level of uniformity before expansion had carried different segments of it too far apart to interact with each other. This is known as the 'horizon problem'. There were other problems, too, but the new inflationary model dealt with them all rather well. Which is why it's now a part of the standard Big Bang model that is generally accepted – despite it having no direct observations to support it, other than the smoothness of the CMBR.

Inflation was an acceleration of space itself, rather than of objects moving through space. So its gravitational-wave signature is not as straightforward as that from conventional accelerating masses – if black holes and neutron stars can ever be described as conventional. In fact, the gravitational waves expected to have been produced during the inflationary period are of such a low frequency that they would show no change during normal human timescales, appearing simply as a frozen pattern imprinted on the CMBR. This pattern is known in the trade as the B-mode polarisation, and may eventually be detected in microwave observations of the CMBR rather than by direct gravitational-wave signals. The bottom line, though, is that the cosmic wallpaper is no barrier to the gravitational signal of inflation. Nor to any of the other physical processes taking place in the Universe's earliest phase.

WHILE THE TWO LIGO DETECTORS THAT MADE THE recent ground-breaking discoveries are amazing in their sensitivity, they are nowhere near sensitive enough to pick up cosmic inflation. Nor are they tuned to the low-frequency gravitational waveband that is necessary to detect events in the immediate aftermath of the Big Bang. However, their descendants almost certainly will be. Today's gravitational-wave technology is still in its infancy, in both design and implementation. More improvements are planned to the LIGO detectors, and there will be more of them. Eventually, it is expected that there will be a network of LIGO-like detectors all around the globe, combining their results to give us a high-frequency gravitational-wave detector the size of Earth.

And beyond that is a space-based detector of even more exquisite sensitivity, which the European Space Agency (ESA) proposes to launch in around 2034. The Laser Interferometer Space Antenna – LISA – will bounce laser beams backwards and forwards over millions of kilometres rather than the 4-kilometre beams of LIGO. It will be sensitive to exactly the low-frequency signals we have just been discussing, with the added possibility of detecting distant supermassive black-hole mergers and the detailed mechanics of galaxy formation. In December 2015, ESA launched a technology demonstrator spacecraft called *LISA Pathfinder*, whose 16-month mission exceeded all expectations. As a proof-of-concept, *LISA*

Pathfinder has succeeded with flying colours, inspiring confidence in the prospects for LISA itself.

With such improvements in technology, there is real hope that we may be able to use gravitational waves to probe the secrets of the early Universe. It might not be too much to hope that one day, we will not only know the physical details of dark energy and cosmic inflation, but the mechanism of the Big Bang itself. And what an astonishing discovery that would be.

CHAPTER 21
UNREQUITED LOVE:
IS ANYONE THERE?

I want to end this book with a good old-fashioned romance. Maybe even a tear-jerker. You might wonder what romance is doing in a science book, but this is a story that plays directly to our emotions, and our sense of who we are. Like all romances, it has protagonists. And, in this case, the protagonists are us and them. Us because we're human, and them – whoever or whatever they are – because we can't stop thinking about them.

It was back in the time of Galileo and Kepler that people first thought seriously about intelligent life in space, and we haven't stopped since. Johannes Kepler was the German mathematician and astronomer who worked out the laws of planetary motion in the early 17th century. In 1610, he wrote a long letter to Galileo in praise of his newly published *Starry Messenger* and the discoveries it reported, particularly in regard to our Moon and the moons of Jupiter. It contains several allusions to the citizens of the Moon (who, he was sure, had created circular embankments to protect themselves from the Sun's radiation) and the inhabitants of Jupiter.

His logic regarding the Jovians is impeccable:

The conclusion is quite clear. Our Moon exists for us on the Earth, not for the other globes. Those four little moons exist for Jupiter, not for us. Each planet, in turn, together with its occupants, is served by its own satellites. From this line of reason we deduce with the highest degree of probability that Jupiter is inhabited.

QED – but I'm not sure that it stands up terribly well. It's on a par with *Star Wars* or the adventures of my childhood space hero, Dan Dare, in Britain's *Eagle* comic. At least their exploits were unashamedly fictional – although Dan Dare's gifted artist, Frank Hampson, threw in a healthy dose of real science, too. Perhaps that's why I'm still excited by this stuff, 60 years later.

Science fiction has imagined extraterrestrial life-forms in every permutation from the hostile to the benevolent. But as Stephen Hawking noted in 2016, 'Meeting an advanced civilization could be like Native Americans encountering Columbus. That didn't turn out so well.' We in Australia don't have to look far to see truly devastating parallels in our own history. And is it even possible that the marauding extraterrestrials might only be interested in trying out a scrumptious new protein source, oven-ready and tastefully dehaired? Yes, it probably is.

That said, there's really no point in trying to hide from them. While we haven't made a habit of aiming radio signals at likely-looking solar systems (apart from a couple of early experiments), our planet has been radio-loud for over 80 years, emitting broadcasts and communications to the Universe at large. Astronomer Seth Shostak of the SETI Institute speaks of 'leakage wafting skywards', which any extraterrestrial society capable of threatening us would be able to detect. And there are five NASA spacecraft leaving the Solar System altogether, to wander through interstellar space for perhaps billions of years. They are the two *Pioneers* (launched in 1972 and 1973), the two *Voyagers* (both launched in 1977) and *New Horizons* (launched in 2006). Each carries tokens of humanity, including directions on how to find us and, in one instance, depictions of how tasty we look. On the issue of beaming signals to advertise our presence to interstellar targets, however, the Breakthrough Initiatives founded in 2015 by Russian entrepreneur Yuri Milner include something called 'Breakthrough Message'. Its declared aim is 'To encourage global discussion on the ethical and philosophical issues of sending messages into space.'

SO, WHERE ARE WE IN THE QUEST TO FIND EXTRATERRES-trial intelligence? The science of astrobiology – the study of the origin, evolution and distribution of life throughout the Universe – is thriving. It is teaching us a lot about ourselves and our fellow earthly species, as well as giving us insights into the possibilities of life elsewhere in the cosmos. We're encouraged by the fact that on Earth, microbial organisms occupy every possible niche, from mountain-tops to deep oceans – not to mention in the planet's crust and atmosphere. And extremophiles such as tardigrades are great examples of the tenacity of life.

It does raise the question, though, of how life is actually defined, and in a cosmic sense, that's not an easy one to answer. It's no good looking for something like DNA and hoping for the best. Less Earth-specific definitions are required. One I quite like defines a living organism as a self-sustaining, self-replicating entity that is capable of Darwinian evolution. You might think that's not specific enough, though, because it's possible to envisage machines that exhibit those characteristics.

My astrobiology colleagues, Paul Davies (Arizona State University) and Charley Lineweaver (Australian National University), want to espouse even broader definitions of life, however – definitions inspired by some unexpected work carried out by the physicist Erwin Schrödinger in the 1940s. You might remember him from his famous quantum cat, which is simultaneously alive and dead until you have a look in its box, whereupon it is decidedly one or the other.

Schrödinger made a valiant attempt to reduce living organisms to applied physics, but came to the conclusion there must be something else going on, too, because life seems to defy the fundamental laws of thermodynamics. Davies proposes that the 'something else' is information, whether it's encoded in DNA or some more amorphous construct such as networks of chemical reactions. Lineweaver goes further, and looks for what he calls 'far from equilibrium dissipative systems' – things that are out of balance with their surroundings in a chemical or physical sense. The trouble with that definition is that it includes entities we wouldn't normally think of as living, like the turbulent atmospheres of planets and stars. Charley Lineweaver is unfazed by that, citing the fact that it makes people think more carefully about such matters.

I THINK IT'S FAIR TO SAY THAT MOST ASTROBIOLOGISTS have tended to avoid these issues by focusing on life as we know it on Earth – bog-standard carbon-containing water-based life. And, in that regard, the results coming from astronomy and planetary science are wholly encouraging. Water is everywhere – it's the most abundant two-element molecule in the Universe. And it's in plentiful supply in the Solar System. Admittedly, much of it is frozen, in the subsurface soil of Mars, the ice-shells of moons like Enceladus, and the nuclei of comets. Even in liquid form, it's more abundant than you might expect. Jupiter's moon Europa, for example, harbours perhaps twice as much water under its icy crust as there is in Earth's oceans, and Saturn's Titan is thought to have even more. And then there's the carbon, which is found all over the place, often locked up in complex organic molecules.

When you broaden your horizons beyond the Solar System to our Milky Way Galaxy, the picture becomes even more promising. Since the only life we are aware of has evolved on a planet, other planets would seem like a good place to start, and the exoplanet community has been doing a grand job of finding them. Our tally of confirmed exoplanets (planets outside the Solar System) passed 4000 in March 2019, and will eventually be joined by another 2870 candidate planets currently awaiting confirmation. And who knows how many more beyond that? The bottom line is that planets are commonplace – something we didn't know in 1995, when the first exoplanet orbiting a normal star was discovered. Statistically, every star in the Galaxy must have at least one planet.

Something else we didn't know about in 1995 was the enormous variety of planetary systems out there – from hot Jupiters that almost skim the surface of their parent stars to remote objects whose orbits take thousands of years to traverse. They range in size from planets many times larger than Jupiter to worlds little bigger than the Moon, and exhibit an equally spectacular range of environments, from frigid ice-worlds to planets so hot that iron drizzles out of the clouds.

Somewhere in the middle of this glittering variety of planethood are those orbiting within the habitable zone of their parent star – the region where water can exist in liquid form. Comfortingly known as the 'Goldilocks zone', it's where the temperature is not too hot and not too cold, but just right. And Goldilocks-zone objects about the size of Earth have a particular appeal to astrobiologists. The detailed study of such worlds is difficult with the current generation of large optical telescopes, but the new generation of 'extremely large telescopes' with mirrors more than 20 metres in diameter will come online in the mid-2020s, providing new capabilities. In particular, the spectroscopic analysis of exoplanet atmospheres will become routine, allowing us to search for biomarkers – the signatures of chemicals associated with living organisms. If those chemicals also include industrial pollutants that could never be created by natural processes, that would be the discovery of the millennium.

Finally, astrobiologists' optimism comes from the sheer numbers that emerge when you look beyond our Galaxy to the wider Universe. The most recent estimate of the number of galaxies that are observable from Earth is two trillion. Typically we'd expect each to contain 100 billion stars or so, resulting in an estimate for the total number of stars in the observable Universe of 2×10^{23}. You can probably guess what's coming next: how does that stack up against Carl Sagan's famous statement that there are more stars in the Universe than grains of sand on all the beaches of the Earth? A long time ago, I checked his calculation, and he was right. In fact, if you throw in that latest estimate of the number of galaxies, he was more right than he could have known. The stars in the Universe outnumber the grains of sand on all the beaches of not one, but two hundred Earths.

SO, IS LIFE ABUNDANT THROUGHOUT THE UNIVERSE? IT may well be. Simple life, at least – single-celled organisms, or microbes. Green slime, perhaps, or its interplanetary equivalent. But what are the odds of those single-celled micro-organisms evolving into complex life-forms, and ultimately into intelligent life? Once again, we immediately run into the problem of definition. What is intelligent life?

In a 1995 article, Carl Sagan defined intelligent organisms as being the functional equivalent of humans. And astrobiologist Charley Lineweaver points to two possible routes for achieving that. One, known in evolutionary biology as convergent evolution, says that the same capability-enhancing traits can be independently acquired by unrelated species with completely different lineages. The evolution of flight is a good example. But, as Lineweaver points out, several environments on Earth have been isolated from each other by the drift of continental plates for far longer than it took the human brain to achieve its present complexity in Africa – but have not yielded any independent equivalent. So he adopts the opposite view, in common with a number of evolutionary biologists. This is that the evolution of intelligence is a quirk of nature resulting from a rare and probably unrepeatable sequence of events.

If that seems depressing, the broader biological picture offers no comfort. We know that the first microbial life appeared on the infant Earth perhaps four billion years ago and a few hundred million years later the first complex organism emerged. There may have been other varieties, but only one survived. How do we know that? Because all complex life on Earth – known as eukaryotic life – can be traced back genetically to that unique progenitor, known as LUCA (the Last Universal Common Ancestor). So far, we have found no trace of an evolutionary false start that could hint at a second genesis of eukaryotic life on Earth.

Once again, thermodynamics enters the picture here, with a few scientists pointing out that eukaryotes are vastly more energy-hungry than their single-celled ancestors, and perhaps that is why their emergence was a one-off. British biochemist Nick Lane has noted that because of the energy demand, 'there is no inevitable evolutionary trajectory from simple to complex life'. He says complex life is just a fluke. This view is shared by other astrobiologists, who are pessimistic about the development of any multi-celled organisms beyond Earth – let alone higher life-forms and extra-terrestrial intelligence.

That pessimism provides a gloomily convincing explanation for the question posed in 1950 by Italian physicist Enrico Fermi.

It's now known as the 'Fermi Paradox', and asks, 'Where is every-body?' It's based on the Copernican principle that there's nothing special about us, so you'd expect intelligent life to be commonplace. Given the hundreds of billions of stars in our Milky Way Galaxy, and its age of around 12 billion years, even if there is only a small probability that intelligent life has evolved elsewhere, its existence should be evident by now. They'd be everywhere. If you can travel close to the speed of light, interstellar distances are no problem. And even if you can't, there's always the possibility of interstellar voyages incorporating successive generations of travellers. Then there are leaked radio transmissions of the kind that Earth has been emitting for decades. So, yes, intelligent species should be detectable unless they have evolved in such a way as to make their presence invisible. Or unless they've been and gone, and are now all extinct. But those thermodynamically minded biologists think it's much more likely that they haven't turned up yet. Except here on Earth.

This view is also supported by a recent Oxford University study that looks in detail at the famous Drake Equation, formu-lated by American astronomer Frank Drake in 1961. The equa-tion attempts to estimate the number of intelligent civilisations in our Galaxy by looking at a series of factors such as the rate at which suitable stars form, the fraction of those stars with plan-ets, the number of those planets suitable for life and the number on which life actually appears. Then you factor in the fraction of life-bearing planets on which intelligence emerges, the number of those that produce technology capable of emitting signals into space, and the fraction that actually go ahead and do so. Most of these factors are just guesses, although at least we now know that most stars do have planets. But with the very best current esti-mates, the new study indicates that there are unlikely to be any other civilisations within the observable Universe. So – the Fermi Paradox is no longer a paradox. They just aren't there.

Should we stop looking for them? No, because of what we might discover on the way – and that's why I'm an enthusiast of initiatives like SETI and Breakthrough Listen. Our quest for higher life-forms inevitably takes us beyond the confines of the Solar System, and we rely on technology that is the stock-in-trade of astronomical research. That means telescopes – optical and radio – with the associated smart technology that we've met elsewhere in this book. In most cases, the technology is the same whether you're investigating the snacking habits of supermassive black holes or seeking signs of intelligent aliens on distant planets. And astrobiology, like most other branches of astronomy, pushes these technologies to their limits.

I PROMISED YOU ROMANCE AT THE START OF THIS CHAPTER. We're still in love with the idea of beings like ourselves going about their business in a galaxy far away. That remains a possibility, of course – and we're never going to be able to prove it's not the case, with all those stars to check out. Until we have evidence for their existence, however, the extraterrestrials will remain in the realm of fantasy. Of course, that does bring its own silver lining. If they don't exist, they can't eat us.

I think there's also a perverse romance in the idea of this vast, incredible Universe that you've been reading about containing only one species able to contemplate it. It is strangely disturbing. What's it all for? Does it suggest that we – as a bizarre and unlikely outcome of the laws of physics and natural selection – don't actually belong here? And, if we weren't here, would the Universe still be?

These are profound questions that we may never be able to answer because we're simply not capable of it. Perhaps the situation was best summed up by the great theoretical physicist and

Nobel Laureate, Max Planck. 'Science cannot solve the ultimate mystery of Nature,' he once remarked, 'and it is because in the last analysis, we ourselves are part of the mystery we are trying to solve.'

★

ACKNOWLEDGMENTS

This book owes a great deal to many people, most of whom are probably unaware of the debt. The ones who aren't are my nearest and dearest, of course, although they're all far too polite to mention it. But I couldn't do this stuff without them. Wholehearted thanks go to my talented and ever-supportive partner in crime, Marnie; my daughters Helen and Anna and their kids Ciarán, Alex, Eve and Hayden (not to mention their partners Liam and Brett), my sons James and Will and Will's partner, Mairead, my two brothers John and Dave and their families, and everyone else who puts up with my random comings and goings. Note to self – tell them all more often how much I love and appreciate them.

It's always a pleasure to acknowledge the encouragement of friends and associates in the world of astronomy, both professional and amateur. In particular, my colleagues at Australia's Department of Industry, Innovation and Science and the former Australian Astronomical Observatory still generously support me in a job that has long filled me with delight.

I also acknowledge the contribution of many other specialists from whom I learn something new every day – about space science, astronautics, Earth and life sciences, science history, mathematics, technology and so on. Thanks, too, to many friends beyond the scientific world whose company is always refreshing – especially the musicians, the travellers and the train buffs. You know who you are.

Then there are the broadcasters. It's been an honour to work with an inspiring group of professionals over more than two decades, mostly at the Australian Broadcasting Corporation. Special mention must go to Philip Clark (who jests that he 'discovered' me in 1997), Chris Bath, Steve Martin and my co-conspirator in the weekly *Space Nuts* podcast, Andrew Dunkley – a former ABC presenter. And, of course, the debonair Richard Glover, whose 'self-improvement' segments on ABC Radio Sydney provided the impetus for this book.

Speaking of which, it's a pleasure to thank the team at NewSouth Publishing who have brought it to life – Kathy Bail, Elspeth Menzies, Paul O'Beirne, Josephine Pajor-Markus, Joumana Awad, Harriet McInerney and others behind the scenes. Not to mention the world's best copyeditor, Jocelyn Hungerford and indexer, Jenny Browne. And thanks, too, to Cathy Axford of Marnie's Dark Sky Traveller company.

Finally, a special word of thanks to the handful of people who encouraged me to pick up my pens and pencils again after a break of more than 50 years – especially artist Dianne Pratt, who kindly provided new ones. Rediscovering the art of sketching for this book was an unexpected pleasure.

INDEX

Illustrations within the text are indicated by *italic* page numbers. Illustrations from the photo section are indicated by *ill*.

1999 NC₄₃ asteroid 44
3200 Phaethon asteroid 37
5691 Fredwatson asteroid 31
67943 Duende asteroid 65

ABC see *Stargazing Live*
Abell 370 galaxy cluster *ill*
active galactic nuclei 188, 195–96
Adams, John Couch 145
Advanced LIGO 212–15, 224
Advanced Virgo 215
Agenzia Spaziale Italiana see *Cassini* spacecraft
Airbus Defence and Space company 59
Airy, George 145
Algol star system 158
aliens see astrobiology
Allan Hills Meteorite 41–42
Allen, David 174–76
amateur astronomy see citizen science
America 35 see also United States
American Association of Variable Star Observers 25
American Journal of Science and Arts 33
Anderson, Eric 64
Anderson, Ryan 105
Anglo–Australian Observatory viii see also Siding Spring Observatory
Anglo–Australian Telescope

10–11, 159–60, 163, 174, *ill*
ANU (Australian National University) ix, 30, 205, 209, 228
Apache Tribal Council 97
Apollo program 50, 66, 71, 79, 99–100
Arago, François 146
Archimedes Institute 56–57
Arecibo, US 182
Aristotelian view 85–91
Arizona, US 97, 119
Arizona State University 228
Arkyd Astronautics 64–65
Asher, David 36
ASKAP (Australian Square Kilometre Array Pathfinder) 52, 183–84, *ill*
asteroids see also meteoric material
   5691 Fredwatson 31
   dislodging meteorites 41
   becoming moons 73
   monitoring 45–46
   ownership and mining of 56–57, 63–68
astrobiology 228–31 see also SETI
   defining life 228–29
   habitability of Mars 104–6, 109–16, 118–23
   habitability of moons 99–101, 107–8, 140
   habitability of space 101–4
Astronautical Federation Congress 103
astronomers, role of 47–48
Astrophysical Journal Letters 186
Astrophysical Observatory, Potsdam 158
Australia and the VLT 184–85
Australian Astronomical

Observatory viii see also Siding Spring Observatory
Australian Astronomical Optics ix
Australian Broadcasting Corporation see *Stargazing Live*
Australian National University ix, 30, 205, 209, 228
Australian Square Kilometre Array Pathfinder 52, 183–84, *ill*
axial tilt, climate and 6

bacteria see astrobiology
Banks, Kirsten 48
Barberini, Maffeo 87–89
Barish, Barry 215
Barry, Trevor 134
Batygin, Konstantin 143–44
BBC see *Stargazing Live*
Bean, Alan 99–100
Beer, UK 101
Bekenstein, Jacob 190, 199
Bellarmine, Robert 86–87
Belt of Venus 17, *18, ill*
Benedix, Gretchen 30
Berenice's Hair 205
Berlin Observatory 145, 201
Bezos, Jeff 60, 62, 122–23
Big Bang
   CMBR and 217–20
   definition 216–17
   expansion and 217, 223
   formation of elements 168
   name 169
   structure imprinted by 160, 210
big crunch hypothesis 221
big splat hypothesis 80
Bigelow Aerospace 68
binary stars 158, 178

Biopan-6 experiment 102
Biosphere-2 facility 119
bipolar nebulae 176
Birkeland, Kristian 20
black holes
    Cygnus X-1 194–95
    evaporation of 198
    formation of 193–94
    gravitational waves and
      214–16
    in Messier 87: 188–89
    name 190
    'no hair' theorem 190
    prediction of 190–91
    Sagittarius A* 196
    Schwarzschild radius 191–92
    singularity 198–99
    sizes of 195–97
Bland, Phil 30
Blue Origin 62–63, 122–23
B-mode polarisation 222–24
Bode, Johann Elert 145
bolides 42–44
Borghese, Camillo (Pope
    Paul V) 86
Bouman, Katie 189
Bowen, Ira Sprague 165–66
Bradford Space 64
Brahe, Tycho 178
Branson, Richard 61–62
Breakthrough Foundation 27,
    228
Britain *see* United Kingdom
British Association for the
    Advancement of Science
    153–54
British Broadcasting Corporation
    see *Stargazing Live*
British Columbia, Canada 185
Brown, Mike 143–44, 149
Bruno, Giordano 85
Bunsen, Robert 154–55, *ill*
Burnell, Jocelyn Bell 194
Burney, Venetia 147

Caccini, Tommaso 86
California Institute of
    Technology 143–44, 149,
    165, 185
Callisto 108
Cambridge Observatory 145
Cambridge University 199
Camden, Lord 94
Canadian Hydrogen Intensity

Mapping Experiment 185
Canberra, ACT *ill*
Cape Canaveral, US *ill*
carbon 102–3, 229
carbon dioxide 7, *8*, 9
Carnegie Institution 206
Cassini spacecraft *ill*
    end of mission 127, 132
    exploring moons 107–8, 132,
      138–40
    exploring Saturn 127,
      129–34, 136
    scientists working with
      133–34
Catholic Church 85–91
Centaurs 128
CERN 211
Cerro Paranal, Chile 184, *ill*
Challis, James 145
Chang'e 4 spacecraft 78
Chapelain, Jean 125
Chariklo 128
Charon 74, 148
Chelyabinsk, Russia 42–44
Chicxulub crater 45
Chile 96
CHIME (Canadian Hydrogen
    Intensity Mapping
    Experiment) 185
China 78, 119, 179
Chiron 128
chondrites 40, 43
citizen science 24–26, 31–32
    Fireballs in the Sky 30
    Galaxy Zoo project 27–29
    observing variable stars 25
    searching for planets 23–24,
      30–31
    SETI projects 26–27
    Stardust@home 27–29
climate change 9
cloud colours 14
Clube, Victor 45
clusters of galaxies 204–5, *208*,
    209, *ill*
CMBR *see* Cosmic Microwave
    Background Radiation
Cold War 58
Colless, Matthew 209
Coma Berenices constellation 205
comets 34–37
commercial enterprise 56–60, *ill*
    see also mining; space tourism
Commercial Space Launch

Competitiveness Act 2015
    (US) 67
Committee on Space Research
    104–6
Commonwealth Scientific
    and Industrial Research
    Organisation 49–50
Comte, Auguste 153–54
Congregation for the Doctrine of
    the Faith 86–90
Connerney, Jack 129
Conrad, Pete 99–100
ConsenSys company 64
conservation 97
Constellation program 58–59
contamination 103–8
Copernicanism 84–90, 167
Córdova, France A 189
Cosmic Microwave Background
    Radiation *ill*
    inflation theory and 223–24
    mapping 219–21
    origin of 217–20
cosmic rays 7
cosmological constant 221–22
cosmological redshift 160
Cox, Brian 29, *ill*
Crab Nebula 179
crepuscular rays 15, *ill*
Crew Dragon 59
CSIRO (Commonwealth
    Scientific and Industrial
    Research Organisation) 49–50
Curiosity 111, 113, *ill*
Curtin University 30
Cygnus X-1 194–95

Daphnis (moon) *ill*
'dark ages' 53–54
dark energy 221–22
dark matter
    alternatives to 206–7
    composition of 207, 209–11
    discovery of 205–6
    mapping *208*, 209–10
    ratio to normal matter 210,
      222
    temperature in early
      Universe and 54
Darmaraland, Namibia 4–5, 9
Darwin, George 72–75
Davies, Paul 228
Deep Space Climate Observatory
    ill

Deep Space Industries 64–65
density waves 170, *ill*
Desert Fireball Network 30
Dialogue by Galileo Galilei on the two Chief World Systems 88–90
Diamandis, Peter 64
Dicke, Robert 190
Dingo, Ernie 51
dinosaurs 45, 131
disc galaxies 205
Discourses and Mathematical Demonstrations Relating to Two New Sciences 91
Doeleman, Shep 189
Dollond, John 93–94
Doppler effect 157–59
Dover Heights, NSW 49
Dragon capsule 59
Drake Equation 233
DSCOVR (Deep Space Climate Observatory) *ill*
Dublin, Ireland 153–54
Dunsink Observatory 96

Earth
  atmosphere and sky 6–7, *8*, 9–19, *18*, 19–21, 114
  geology 4–7, *8*, 9
  life on 6, 115–18, 229, 232
  revolution around Sun 85–91, 167
  role in formation of moon 72–77, 80
Earthshine 6
Eddington, Arthur 201
EDGES (Experiment to Detect the Global EoR Signature) 53–55
Egypt 40–41
Einstein, Albert 146, 200–202, 222 *see also* General Theory of Relativity
Einstein rings and crosses 202–3, *ill*
El Tatio, Chile 122
electromagnetic radiation 48–49, 53 *see also* Cosmic Microwave Background Radiation; telescopes
elements 168–69 *see also* mining
Enceladus (moon) 107–8, *ill*
Enterprise 62
Epimetheus (moon) 126
Eros (asteroid) 56–57

ESA *see* European Space Agency
ESO *see* European Southern Observatory
ESO 601-G036 galaxy 184
Eta Carinae star system 178
Europa (moon) 107–8
Europe 102
European Nuclear Research Centre 211
European Southern Observatory 96, 176, 184–85, *ill*
European Space Agency 108, 114, 121–22, 224, *ill* see also *Cassini* spacecraft; Hubble Space Telescope
Event Horizon Telescope 188, 196
evolution 231–32
ExoMars 121–22
Experiment to Detect the Global EoR Signature 53–55
extraterrestrial life *see* astrobiology
extreme trans-Neptunian objects 142–43

Falcon vehicles 59, 119, *ill*
Fast Radio Bursts
  detection of 180–85
  possible causes 182–87
Fermi Paradox 232–33
fibre optics 162, 222
Finocchiaro, Maurice 88–89
fireballs 39, 42–44
Fireballs in the Sky project 30
fission theory of Moon's origin 72–73
Flagstaff, US 147
Florence, Italy 86, 90
'flywheel' effect of Moon 6
Ford, Kent 206
fossil fuels 9
Frankland, Edward 164
FRBs *see* Fast Radio Bursts
Freeman, Ken 205
Freundlich, Erwin 201–2
Friedmann, Alexander 217

GALAH (Galactic Archaeology with HERMES) 163
galaxies *ill* see also Large Magellanic Cloud; spiral galaxies
  black holes in 188, 195–96

dark matter in 204–7, *208*, 209
  number of 231
galaxy clusters 204–5, *208*, 209, *ill*
Galaxy Zoo project 27–29
Gale Crater 111, *ill*
Galileo
  Copernicanism 83–91
  Saturn and 124
Galileo probe 107–8
Galle, Johann Gottfried 145
Ganymede (moon) 108
Geminid meteors 37
gender in astronomy 47–50, 206
General Theory of Relativity 146, 200–204, 213 *see also* gravity
giant impact hypothesis 75–77, 80
global warming 9
Goddard Space Flight Center 129
Goldilocks zone 230
Gondwana 4
gravitational lensing 202–4, 209, *ill*
gravitational waves
  detection of 212–16
  inflation theory and 222–24
  origin of 213–15
gravity 221–22 *see also* black holes; dark matter; General Theory of Relativity; Theory of Universal Gravitation
Green Bank Observatory 27
green flash 15–16
green pea galaxies 28
greenhouse effect 7, *8*, 9, 114
Grey, Andrew 23–24
Gump, David 64
Gusev Crater 122
GWs *see* gravitational waves

Halley's Comet 37
Hanford, US 214
Hanny's Voorwerp 28
Harvard-Smithsonian Center for Astrophysics 186
Harvey-Smith, Lisa 48
Haumea (dwarf planet) 128
Hawaii 97–98, 129, 178, *ill*
Hawking, Stephen 198–99, 227
Hawking radiation 198

helium 164
HERMES (High Efficiency and
    Resolution Multi-Element
    Spectrograph) 163
Herschel, William 145
Hill, John 162
Holy Roman Inquisition 86–90
Hooke, Robert 92
Hope, Dennis M 56
horizon problem 223
Hoyle, Fred 102, 168–69
Hubble flow 159–61, 220–21,
    223–24
Hubble Space Telescope 176–77,
    209, ill
Huggins, William and Margaret
    156–57, 163–64
Huygens 138
Huygens, Christiaan 125, 137
hydrogen 26, 53, 156, 168, 210

Icelandic volcanism 5
inflation theory 222–24
infrared telescopes 97, 129,
    176–77, 209, ill see also Very
    Large Telescope
Inquisition 86–90
Inspiration Mars Foundation 69
interferometers 212–15, 224–25
International Astronomical
    Union 83, 142
International Space Station 58,
    60, 101
International Year of Astronomy
    83
interplanetary dust 21–22, 38
interstellar medium 169
iron 40–41
Israel's Weizmann Institute 206
Istoria e Dimostrazioni intorno
    alle Macchie Solari 85–86,
    88, 124
Italian Space Agency see Cassini
    spacecraft

Japan 76, 178
Jesuits 85–86
Jet Propulsion Laboratory 133
Jodrell Bank Radio Observatory
    51
Jupiter
    'Jovians' 226–27
    moons of 84–86, 107–8
    rings of 128–30

Jupiter Icy Moons Explorer 108

Kamchatka peninsula 45
Karbysheva, Yulia 43
Karoo Array Telescope 51–52
Keck telescopes 129
Keeler, James 126
Keeler Gap ill
Kepler, Johannes 173, 226
Kepler spacecraft 23, 31
Kirchhoff, Gustav 154–55, ill
Klavans, Val, picture of Venus ill
Kraken Mare 139–40
Kuiper Belt 73

Lake Chebarkul 43
Lambda CDM model see Big
    Bang; dark energy; dark matter
Lane, Nick 232
Large Binocular Telescope
    96–97
Large Hadron Collider 211
Large Magellanic Cloud 173–76
Las Cumbres Observatory 31
laser interferometry 212–15,
    224–25
Le Verrier, Urbain Jean-Joseph
    145–46
Lemaître, Georges 217
Leonid meteors 33–36
Letters on Sunspots 85–86, 88,
    124
Lewicki, Chris 64–65
life beyond Earth see
    astrobiology
Ligeia Mare 139
light echoes 170–71, 172, 173
    of historical supernovae
        178–79
    of Supernova 1987a 173–76
    of V838 Monocerotis 176–77
light pollution 54
LIGO (Laser Interferometer
    Gravitational-Wave
    Observatory) 212–15, 224
Lineweaver, Charley 228–29,
    231–32
Lingam, Manasvi 186
Lintott, Chris 28, 31
Lipperhey, Hans 84
LISA (Laser Interferometer
    Space Antenna) 224–25
Livingston, US 214
Lockyer, Norman 164

Lodge, Oliver 202–3
Loeb, Avi 186
Lorimer Burst 180–82
Lowell, Percival 109, 147
Lunar Embassy 56
Luxembourg's space law 67
Lynx rocket 62

MACHOs (Massive Compact
    Halo Objects) 207
magnetosphere 6–7, 114
Malin, David 174
Mariner missions 109
Mars ill
    conditions and habitability
        104–6, 109–16, 118–23
    meteors and 40–41
    travel to 58–59, 68–69
Mars Express orbiter 114, 121, ill
Mars Global Surveyor 112
Mars One 120
Mars2020 122
Massive Compact Halo Objects
    207
Maui ill
Mauna Kea 97–98, 129, ill
Maxwell, James Clerk 126
May, Brian 21
McNaught, Robert 36, 173
MeerKAT 51–52
Mercury 146
Messier 87 188–89
meteoric material ill see also
    asteroids; fireballs
    ALH84001 41–42
    Fireballs in the Sky 30
    Geminids 37
    Leonids 33–36
    makeup of 40–42
    Orionids 36
    sources of 34–36, 40–41, 71
    stargazing for 37–39
    terminology 34, 39–40
methane on Mars 121
Michell, John 190–91
microbes see astrobiology
microwaves see Cosmic
    Microwave Background
    Radiation
Mid-Atlantic Ridge 5
Milgrom, Mordehai 206–7
Milky Way 4, 20, 167, 195–97
Miller, William 156
Milner, Yuri 27, 228

mining 57, 63–69
Modified Newtonian Dynamics 207
MOND (Modified Newtonian Dynamics) 207
Moon 79, *ill see also* Apollo program
  eclipses vii
  effect on Earth 6, 77
  Johannes Kepler on 226
  light on 13
  lunar dichotomy 77–78, 80
  meteors and 40–41, 71
  orbit 77–78, 167
  origin of 72–77
  ownership and mining of 56, 69
  surface of 78, 80
  TLPs 71
  travel to 58, 68
moons 73–74, 84, 107–8, 126, *ill*
Moor Hall, Chester 93–94
Moore, Patrick vii–viii, 71
Moses, Beth 62
Mount Graham, US 97
multi-fibre spectroscopy 162, 222
Murchison Radio-Astronomy Observatory 51–55, *ill see also* Australian Square Kilometre Array Pathfinder
Musk, Elon 59, 118–20

Napier, Bill 45
Narrabeen Lagoon, NSW *ill*
NASA (National Aeronautics and Space Administration) *see also* Apollo program; *Cassini* spacecraft; Hubble Space Telescope; *Voyager* missions
  Constellation program 58–59
  Galileo 107–8
  Kepler 23, 31
  NEAR-Shoemaker probe 57
  New Horizons 148, 227
  on Mars 111–14, 122, *ill*
  Pioneers 126–27, 227
  Space Shuttle program 60
  Stardust@home 28
Nature 102
Neanderthals 117
NEAR-Shoemaker probe 57
nebulae 163–66, 170–71 *see also* reflection nebulae

Nemitz, Gregory W 56–57
Neptune 128–30, 145, 147–48
neutron stars 180, 194, 216
New Horizons 148, 227
New Zealand astronomy 25
Newton, Isaac 93, 144–45, 153, 200 *see also* Theory of Universal Gravitation
NGC 2044 star cluster 175
Northrop Grumman company 59, 61

Obama, Barack 58
occultation 128
oceans 6, *8*, 9, 77
Olmsted, Denison 33–34
Olympus Mons 110
On the Stability of the Motion of Saturn's Rings 126
Open University 101
Ophiuchus constellation 173
Opportunity 111
optical telescopes *see also* Anglo–Australian Telescope; United Kingdom Schmidt Telescope; Very Large Telescope
  Galileo and 83–84, 124
  importance of 48
  Keck telescopes 129
  Large Binocular Telescope 96–97
  Royal Astronomical Society and 95–96
  sizes 96–97, 230
  SkyMapper telescope 30
  Subaru Telescope 178
  technology 83–84, 91–95, 161–63
  Thirty Meter Telescope 97
Orionid meteors 36
Outer Space Treaty 66–67, 103–4
Owens Valley Radio Observatory 185
Oxford University 233
oxygen 165

Pandora (moon) *ill*
panspermia 102–3
Pan-STARRS1 *ill*
Paris Observatory 145
Parkes Observatory 27, 50, 180–82
particle physics 210–11
Paul V, Pope 86

Payne-Scott, Ruby 49–50
Perlmutter, Saul 221
Philosophical Transactions of the Royal Society 190–91
Phoenix lander 114
Pioneer missions 126, 227
Pisa, Italy 215
Planck, Max 235
Planet Nine 142–44, 148–50
  on *Stargazing Live* 30–31
Planetary Resources company 64–65
planetesimals 40–41
planets 23–24, 30–31, 230 *see also names of specific planets*
plate tectonics 4–5, 7, *8*, 9, 113–14
Pluto 142–43, 147–48
Porco, Carolyn 133–34
Potsdam, Germany 158
Ptolemaic view 85–91
pulsars 180, 194, 216 *see also* Fast Radio Bursts
Punga Mare 139

quantum theory 165, 198–99
quasar activity 197, 203, *ill*

Radial Velocity Experiment 163
radio astronomy *see also* Event Horizon Telescope; Fast Radio Bursts; Murchison Radio-Astronomy Observatory; Parkes Observatory
  CMBR and 217–19
  development of 49–50
  hydrogen in 26
  MeerKAT 51–52
  SETI and 26–27
Ramsay, William 164
RAVE (Radial Velocity Experiment) 163
Reeve, Richard 91–93
reflection nebulae *see* light echoes
renewable energy 9
Rhea (moon) *ill*
Riess, Adam 221
rings of celestial bodies 124–33
Rome, Italy 86
Roscosmos 60, 122
rotation curves 206
Royal Astronomical Society 95–96

INDEX

Royal Observatory, Edinburgh 45
Rubin, Vera 206
Ruby Payne-Scott award 50
Russia 42–45, 58
Russian Space Agency 60, 122
Rutan, Burt 61
Ryder, Stuart 185

Sagan, Carl 138, 231
Sagittarius A* 196
San Carlos Apache people 97
Saturn
    atmosphere and weather 134, *135*, 136, *ill*
    moons of 107–8, 126–27, 136–41, *ill*
    rings of 124–27, 129–33
Scaled Composites company 61–62
Schawinski, Kevin 24, 28
Scheiner, Christoph 85–86
Schiaparelli, Giovanni 109
Schmidt, Brian 221
Schneider, Étienne 67
Schrödinger, Erwin 228–29
Schwarzschild, Karl 191–93
Schweickart, Rusty 119
search for extraterrestrial intelligence *see* SETI
seasons 6
Sedna (dwarf planet) 143
SELFI (Submillimetre Enceladus Life Fundamentals Instrument) 108
Sentinel *ill*
SETI
    arguments for and against 227–28, 234
    citizen science 26–27
    FRBs and 186–87
    Johannes Kepler on 226–27
    odds of intelligent life in space 231–33
Sheepshanks, Richard 95–96
Sheppard, Scott 143
shooting stars *see* meteoric material
Shostak, Seth 227
Siberian superbolide 42
Sidereus nuncius 84
Siding Spring Observatory viii–ix, 10–11, 36, 54 *see also* Anglo–Australian Telescope;

SkyMapper telescope; United Kingdom Schmidt Telescope
Simonyi, Charles 60
SKA (Square Kilometre Array) 51–52
Skylon rocket 68
SkyMapper telescope 30
Slebarski, Tadeusz 202
Slipher, Vesto 161
Solar System
    Copernican vs Aristotelian view 84–91
    formation of 103
solar wind 7, 114
South, James 95
South African radio astronomy 51–52
South Pole–Aitken basin 78, 79
Soviet Union *see* Russia
Soyuz vehicles 59–60
Space Adventures company 60–61
Space Industry Act 2018 (UK) 63
space law 63, 66–67, 103–4
Space Race 58
Space Shuttle program 60
space tourism 60–61, 68–69
Spaceport America 62
SpaceShipOne 62
SpaceX 59, 118–20, *ill*
spectroscopy
    cosmological redshift 160
    dark matter and 205–6
    development of 153–55
    Doppler effect 157–59
    radio 180–81
    spectra of atmospheres 230
    spectra of nebulae 164–66
    spectra of stars 155–56, 164
    technology 156–57, 161–63
    Zeeman effect 159
Spilker, Linda 133
spiral galaxies 170, 184–85, 204–5 *see also* Milky Way
Spirit 122
Square Kilometre Array 51–52
Standard Model of Particle Physics 210–11
Stardust@home 28
Stargazing Live 23–24, 29–31, *ill*
Starry Messenger 84
stars 53–54, 153, 156, 178, 231, *ill see also* neutron stars; supernovae; V838 Monocerotis;

white dwarfs
Starship 59, 119
steady-state theory 169, 217
Streptococcus mitis 100
Subaru Telescope 178
Submillimetre Enceladus Life Fundamentals Instrument 108
Sun 7, 13–14, 114, 155–56 *see also* Copernicanism
sunset *see* twilight
supernovae
    light echoes 170–71, *172*, 173–76, 178–79
    process of explosion 168–70
Surveyor 3: 99–100
Swinburne University 134
Sydney, NSW *ill*
Systema Saturnium 125

tardigrades 102, *ill*
TD1A spacecraft viii
tectonic movement 4–5, 7, 8, 9, 113–14
telescopes 45–49, 194 *see also* infrared telescopes; optical telescopes; radio astronomy; ultra-violet radiation
Tempel, Wilhelm 35
terminator line 11–12, 14
Tesla vehicles 59
Tharsis Rise 110
Theia theory 75–77
Theory of Universal Gravitation 144–46, 200 *see also* gravity
Thirty Meter Telescope 97
Thorne, Kip 215
Thornhill, UK 190
Throat of Kraken 140
tides 6, 77, 88
Titan (moon) 107, 137–41, *ill*
Tito, Dennis 60, 69
Toba supervolcano 117
Tombaugh, Clyde 147
transient lunar phenomena 71
transits 23–24
Troughton, Edward 95
Trujillo, Chad 143
Truth or Consequences, US 62
Tumlinson, Rick 64
Tunguska superbolide 42
Turner, Herbert Hall 147
Tutankhamun 41
Tuttle, Horace Parnell 35
twilight

stages of 19
terminator 11–12
visual phenomena 14–17, *18*,
19–21, *ill*

ultra-violet radiation 48, 53, 137
telescopes viii, 176–77,
209, *ill*
United Kingdom viii, 63, 68,
101, 145, 201
United Kingdom Schmidt
Telescope 161, 163
United Launch Alliance 59
United Nations Outer Space
Treaty 66–67, 103–4
United States *see also* NASA
commercial enterprise in
space 58, 63, 66–67
EDGES 53–55
Leonid meteor showers
33–36
LIGO 212–15, 224
research on Theia 76
Space Race 58
Unity 62
Universe
creation theories 169, 217 *see
also* Big Bang
expansion of 159–61, 220–
21, 223–24

processes in 167–68
typical place in 3
University of Arizona 96–97
University of Buckingham 102
University of Cambridge 72
University of Dublin 96
University of Hawaii 97–98
University of Heidelberg 154
University of Southern
Queensland 159
University of Sydney 49
Uranus 128–30, 145, 147–48
Urban VIII, Pope 87–89
UV *see* ultra-violet radiation

V838 Monocerotis 176–77, *ill*
Venus 19, 84, 88
Very Large Telescope 96, 176,
184, *ill*
Viking missions 109
Virgin Galactic 61–63
Vogel, Hermann Carl 157
volcanism *8*, 9, 71, 110, 117
Voyager missions 127, 129, 138,
148, 227
VSS Enterprise and *Unity* 62

Washington D.C., US 206
Watson, Fred (author) vii–ix, 11,
31, 162

Weakly Interacting Massive
Particles 207, 209
Weiss, Rainer 215
Weizmann Institute 206
West Virginia University 180
Western Australia 51–52, *ill*
Wheatstone, Charles 153–54
Wheeler, John 190
white dwarfs 193
Wickramasinghe, Chandra 102
WIMPs (Weakly Interacting
Massive Particles) 207, 209
Wolf-Rayet stars and nebulae *ill*
women, role in astronomy
47–50, 206
Woolley, Richard 50
Worshipful Company of
Spectacle Makers 94
Wow! signal 27

XCOR Aerospace 62
X-ray astronomy 194

Yale College 33
Yearbook of Astronomy 175
Yuegong-1 119

Zeeman effect 159
zodiacal light 20–22
Zwicky, Fritz 204–5